Farmers' Favourites

Farmers' Favourites

Compiled by Sally Mitchell

Line drawings by Mary Beck

Royal Agricultural Benevolent Institution

Old Pond Publishing

First published 1996
Reprinted 2007

ISBN 978-1-905523-77-1

A catalogue record for this book is available from the British Library

**It is important to decide in advance whether to use
imperial or metric measures and then use only one kind
in the recipe concerned. Do not mix metric and imperial
measures in one recipe.**

Size 3 refers to medium-sized eggs.

**Published by Old Pond Publishing
Dencora Business Centre
36 White House Road, Ipswich IP1 5LT
United Kingdom**

Distributed in North America
by Diamond Farm Enterprises,
RR3 16385 Telephone Rd S
Brighton, ON KOK 1H0
Canada
www.diamondfarm.com

Cover design by Hannah Berridge
Typeset by Galleon Typesetting, Ipswich
Printed and bound by Biddles Ltd, King's Lynn

Contents

Foreword

Lord Plumb DL MEP

This cookery book, compiled for the benefit of the many farming families who are supported by the Royal Agricultural Benevolent Institution, should find a place in every farm kitchen in the country.

There is a wealth of traditional British country food, prepared by cooks from the rural community, which can compete with produce throughout the world, and the recipes listed here make fine dishes for discerning consumers.

Since the many beneficiaries from RABI have subscribed to this book, saying that a recipe is a small price to pay for the help they receive, we can all benefit from the wealth of their experience and share some country fare whilst helping to promote our charity's work.

We are grateful to Sally Mitchell for her initiative and enterprising idea.

Introduction

Here is a collection of farmers' recipes which have been tried and tested and passed down through the family – these always prove the best.

Sally Mitchell sent me advance copies of some of the recipes to go in the book – the delightful part was that most of them came with a personal letter singing the praises of RABI. Several of the recipes are using up gluts from the orchard such as apples, plums and damsons, and one farmer's wife was very keen to include rabbit stew, which to many country folk comes free.

I wish this book every possible success and I hope you enjoy sharing the recipes.

Very best wishes,

Mary Berry

What is the *RABI?*

R·A·B·I
Supporting
Farming Families
Charity No. 208858

The Royal Agricultural Benevolent Institution is the national charity for assisting retired, disabled or other disadvantaged members of the farming community. After a lifetime of hard work on the land, many find they have little or no money left for their retirement years:

- A widow may find herself alone with a rising tide of bills and financial problems and no one to turn to for help.
- A tenant farmer may be forced to leave a home of many years.

- Savings can soon disappear if there are residential or nursing home fees to pay.
- Farming families who are on low incomes may find they cannot cope during times of crisis.

RABI provides regular long-term support to those with limited savings who are fully retired due to age, permanent disability or ill-health, in the form of a quarterly living grant. One-off help is also provided when long-term beneficiaries are unable to pay for essential items or require disability aids.

RABI assists towards the cost of home help, enabling beneficiaries to stay in their own homes for as long as they are able, and with residential care home fees.

In most cases, beneficiaries will have worked full-time in farming for at least 10 years.

Support is also available for working farmers and their families experiencing exceptional difficulties, such as bereavement or ill health in addition to financial problems, and when facing hardship caused by events beyond their control, such as flooding or outbreaks of animal disease.

RABI understands the problems faced by the farming community and can take immediate action to help those in need. We believe it is important that our beneficiaries have the peace of mind of knowing there is somebody there to turn to in times of trouble or need. Help can be just a phone call away – the RABI confidential helpline number is 01865 727888.

None of the support that RABI gives to its beneficiaries would be possible without the generosity of its donors. To find out more about RABI and how you can help its work, visit www.rabi.org.uk

Shaw House, 27 West Way, Oxford OX2 0QH
Tel: 01865 724931   Fax: 01865 202025

Registered charity no. 208858

As a member of the Farming Help Partnership, RABI works closely with Farm Crisis Network and the ARC-Addington Fund to provide a comprehensive range of practical, emotional and financial support for those in need in the farming community.

Preface

As a council member for the Royal Agricultural Benevolent Institution, I am always looking for fund-raising ideas. The idea of a recipe book came from the wealth of traditional country food prepared by cooks from the farming community, and what better source of such recipes than our beneficiaries. Each of our 1000+ beneficiaries received a written request for their favourite recipe, and not surprisingly I was overwhelmed by their saying that a recipe was a small price to pay for all the help they receive from the RABI.

From all parts of the country, old, new and variations of favourite recipes were sent in and all were tested by trained and experienced cooks. The best and more unusual have been selected. I hope you will enjoy them as much as I have in preparing this selection of *Farmers' Favourites*.

I have thoroughly enjoyed putting this collection of recipes together and must thank Mary Beck for her wonderful illustrations, which I think set the scene and complement the theme excellently. This book would not have been ready in time without the hard work of my recipe-testing friends, namely Peggy Copsey, Helen Crowder, Nell Curtis, Phyl Darbishire, Carolyn Gay, Liz Hill, Margot Pickering, Emily Southgate, Sarah Southgate, Bridget Starkings, Kandy Trigg-Dudley and Norma White, with extra special thanks to Carolyn and Kandy for their time and their typing skills.

Lastly I must not forget to thank my husband and three sons for eating their way through the recipe book uncomplainingly.

SALLY MITCHELL

Light Meals

Leek and Sweetcorn Soup

450g (1lb) sweetcorn cobs or large tin (326g)
    sweetcorn
2 leeks
75g (3oz) butter
60ml (4 tbsp) plain flour
600ml (1pt) white stock
600ml (1pt) milk (or dry white wine)
salt, pepper and nutmeg
150ml (5fl oz) cream

If using fresh corn, boil the cobs for 6–7 minutes. Cut the kernels off
the cobs. If using a tin, drain it.

Trim the roots and leafy parts from the leeks; cut them in half
lengthways and wash thoroughly. Chop finely. Melt the butter in a
pan, add the leeks and cook gently for 5 minutes or until soft. Stir in
the flour and cook through before gradually blending in the stock and
milk.

Add the corn kernels, bring to the boil, season to taste and allow to
simmer for 15 minutes. Rub through a sieve or liquidise in a blender.

Return to the pan and reheat. Correct seasoning if necessary. Blend a
little of the hot soup with the cream, stir back into the pan and heat
through without boiling.

Serves 4–6

H. & N. Major
Gloucestershire

Leek and Potato Soup

450g (1lb) potatoes sliced
3 leeks (white part only)
600ml (1pt) marrowbone stock or 1 chicken
    stock cube dissolved in water
50g (2oz) butter
cream and milk as required
salt and pepper to taste
chopped chives and parsley

Clean the leeks, cut into pieces and sauté in half the butter, taking care not to let them brown. Add the potatoes, stock and seasoning. Return to the heat and cook until the vegetables are soft.

Liquidise and return to the pan, adjust the seasoning, add some cream and milk to make required consistency. Reheat but do not boil. Before serving sprinkle with the chives and parsley.

Serves 4

R. Willis
Devon

Celery Soup

50g (2oz) butter
2 medium onions (chopped)
2 medium potatoes (chopped)
2 heads of celery (chopped)
1 clove of garlic (crushed)
300–450ml (½–¾pt) stock
1 small tin evaporated milk
chopped parsley
seasoning

Melt the butter in a large saucepan, add the vegetables and gently cook without browning. Liquidise the cooked vegetables and then add the stock and evaporated milk. Season to taste. Heat through but do not boil. Sprinkle with the chopped parsley and serve with crusty fresh bread.

Serves 4–6

F. Sullivan
Dyfed

White Vegetable Soup

1 carrot
1 leek
1 onion
1 stick celery
1 large potato
1 turnip
2 bay leaves
50g (2oz) butter
600ml (1 pt) white stock or chicken stock cube
   dissolved in water

5ml (1 tsp) sugar
salt and pepper
25g (1oz) cornflour
300ml (½ pt) milk

Cut the vegetables into small pieces. Melt the butter in a saucepan, add the vegetables and cook gently for 5 minutes without browning. Add the bay leaves, stock, sugar and seasoning and simmer for 15 minutes until the vegetables are soft.

Mix the cornflour to a smooth paste with the milk, add to the vegetables and bring to the boil, stirring carefully. Remove the bay leaves before serving.

Serves 4–6   *Powys*

Mutton Broth

25g (1oz) pearl barley
450g (1lb) scrag end of mutton
900ml (1½ pt) cold water
1 carrot (chopped)
1 onion (chopped)
1 small turnip (chopped)
salt

Cut mutton into small pieces, removing any fat. Wash barley and put with mutton and mutton bones into a large saucepan with the water. Bring slowly to the boil, skim, then cover and cook slowly for 2 hours.

Add chopped vegetables, and season to taste. Then cook for a further 1 hour. Remove bones before serving.

Serves 4   *V Shinn*
                                                       Norfolk

Asparagus Flan

PASTRY
>   175g (6oz) plain flour
>   pinch of salt
>   40g (1½oz) butter or hard margarine
>   40g (1½oz) lard
>   75g (3oz) strong cheese (grated)
>   cold water to mix

FILLING
>   450g (1lb) tin asparagus (drained)
>   40g (1½oz) butter
>   40g (1½oz) plain flour
>   450ml (¾pt) milk
>   100g (4oz) strong cheese (grated)
>   salt, cayenne pepper, dry mustard powder
>   chopped parsley

Preheat oven to 190°C / 375°F / Gas mark 5.

Sift the flour and salt into a mixing bowl, and then rub in the fats until mixture resembles breadcrumbs. Add the grated cheese and with a round-bladed knife stir in the cold water to make a firm dough. Knead and roll out. Then line a 20.5–25.5cm (8–10in.) flan tin with the pastry, prick with a fork and bake blind.

When the flan case has cooled, arrange the drained asparagus in the base.

To make the filling, melt the butter and stir in the flour. Gently cook for 1 minute, then add the milk gradually, stirring all the time.

Bring to boiling point. Remove from heat and then stir in 75g (3oz) of the grated cheese and the seasonings.

Pour the sauce over the asparagus and sprinkle with the remaining 25g (1oz) grated cheese. Heat through and brown under the grill. Garnish with chopped parsley.

Serves 8

H. & N. Major
Gloucestershire

Tuna and Cucumber Flan

PASTRY
175g (6oz) plain flour
40g (1½oz) butter or hard margarine
40g (1½oz) lard
pinch of salt
cold water to mix

FILLING
211g (7½oz) can tuna in brine
5cm (2 in) cucumber (diced)
60ml (4 tbsp) thick mayonnaise
150g (5oz) carton soured cream
10ml (2 level tsp) powdered gelatine
30ml (2 tbsp) water
10ml (2 tsp) wine vinegar
salt and pepper
cucumber slices to garnish

Preheat oven to 190°C / 375°F / Gas Mark 5.

Mix flour and salt in a large bowl, then rub in the fats until the mixture resembles fine breadcrumbs. Using a round-bladed knife to mix, add enough cold water to form a stiff dough. Turn dough out onto a floured surface and knead lightly. Roll out and line a 20.5cm (8in.) flan tin. Bake blind and leave to cool.

Drain and flake the tuna into a bowl. Stir in the diced cucumber, mayonnaise and soured cream. Dissolve gelatine in cold water and vinegar in a small bowl over hot water. Cool and then stir into the fish mixture. Season to taste and pour into the pastry case. Chill before serving, garnished with cucumber slices.

Serves 6

P. Frizell
Dorset

Cheese and Carrot Flan

PASTRY
>175g (6oz) plain flour
>40g (1½oz) butter or hard margarine
>40g (1½oz) lard
>pinch of salt
>cold water to mix, approximately 30ml (2 tbsp)

FILLING
>50g (2oz) cheese
>100g (4oz) carrots
>60ml (4 tbsp) milk
>1 size 3 egg
>salt and pepper to taste

Preheat oven to 200°C / 400°F / Gas mark 6.

Mix flour and salt in large bowl, rub in fats to resemble breadcrumbs. Using a round-bladed knife to mix, add cold water to form a stiff dough.

Turn out dough onto floured surface and knead lightly. Roll out and line a 18cm (7in.) flan dish.

Grate the peeled carrots and the cheese into a basin, adding salt and pepper. Beat egg and milk together and add to the carrots and cheese. Pour into pastry case.

Bake for approximately 30 minutes. Serve warm.

serves 6–8

J. Todd
E. Yorkshire

Egg and Salmon Flan

PASTRY
    175g (6oz) plain flour
    40g (1½oz) butter or margarine
    40g (1½oz) lard
    pinch of salt
    cold water to mix

FILLING
    3 size 3 eggs
    150ml (¼pt) milk
    75g (3oz) cheese (grated)
    211g (7½oz) can salmon (drained and flaked)
    salt and pepper
    2–3 tomatoes (sliced)
    chopped parsley

Preheat oven to 190°C / 375°F / Gas mark 5.

Mix flour and salt in a large bowl. Rub in fats until mixture resembles breadcrumbs. Using a round-bladed knife to mix, add the cold water to form a stiff dough. Turn dough onto floured surface, knead lightly and roll out. Line a 20.5cm (8in.) flan tin and bake blind.

Beat the eggs and milk together. Stir in half the cheese, then the drained and flaked salmon, and the salt and pepper. When blended pour into the flan case and sprinkle on the remaining cheese. Bake for 25 minutes, when flan should be golden brown and set. Garnish with tomato slices and sprinkle with chopped parsley. Serve warm or cold.

Serves 4–6   F. Cook
                                                    Merseyside

Sausage and Leek Pie

700g (1½lb) potatoes (peeled and sliced)
25g (1oz) butter
450g (1lb) sausages sliced
1 onion (chopped)
1 clove garlic (chopped)
4 leeks (cleaned and sliced)
40g (1½oz) flour
600ml (1pt) milk
100g (4oz) cheddar cheese (grated)

Preheat oven to 190°C / 375°F / Gas mark 5.

Cook potatoes in boiling salted water until soft (approximately 5–10 minutes). Melt butter in a large frying pan, add sausages and cook for 5 minutes. Remove sausages and put in 1.2 litre (2pt) ovenproof dish. Add onion, garlic and leeks to frying pan and cook until leeks are soft. Add the flour and cook for 1 minute, then gradually add the milk and 75g (3oz) cheese. Bring to the boil and simmer for 1–2 minutes, stirring all the time. Pour over the sausages.

Transfer mixture to dish with sausages. Arrange the potato slices on top, sprinkle cheese on top of the potatoes.

Bake in oven for 20 minutes, until cooked through and the topping is golden brown.

Serves 6

Norfolk

Bacon and Egg Supper

6 rashers back bacon
6 tomatoes (halved)
12 mushrooms (sliced)
6 size 3 eggs
butter

Preheat oven to 200°C / 400°F / Gas mark 6.

Lightly grease 6 large ramekins.

Cut bacon rashers in half and dry fry. Into each ramekin place 2 pieces of bacon, 2 halves of tomatoes and some mushroom slices. Dot with a knob of butter and bake for 10 minutes.

Crack an egg into the centre of each dish and bake for a further 5 minutes.

Serve at once with hot garlic rolls.

Serves 6

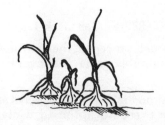

L. Griffiths
Dyfed

Bacon and Potato Rolls

75g (3oz) plain flour
5ml (1 tsp) baking powder
50g (2oz) margarine
pinch of salt
75g (3oz) mashed potatoes
cold water
225g (8oz) streaky bacon

Preheat oven to 200°C / 400°F / Gas mark 6.

Make pastry by rubbing margarine into sieved flour, baking powder and salt until mixture resembles breadcrumbs. Add mashed potatoes and bind together with a little cold water. Roll out thinly on a floured surface.

Lay rashers of bacon on top of pastry, and cut into strips according to the length of the bacon. Roll up and place on a greased baking tray.

Bake for 20–30 minutes.

Serves 4

Hampshire

Cheese Soufflé

50g (2oz) butter
22½ml (1½ tbsp) wholemeal flour
150ml (¼pt) milk
75g (3oz) strong cheese (grated)
3 size 3 eggs (separated)
salt and cayenne pepper

Preheat oven to 180°C / 350°F / Gas mark 4.

Melt the butter in a small saucepan, then stir in the flour and cook over a low heat for a minute or so. Add the milk gradually, stirring all the time.

Remove from the heat and mix in the egg yolks and grated cheese. Season to taste. Allow to cool. Beat the egg whites until stiff and fold into the cooled mixture.

Turn into a well-greased 1.2 litre (2pt) soufflé dish. Bake for 20–30 minutes. This will rise a few centimetres above the top of the dish.

Serve with a mixed salad.

Serves 2

K. Ainsly
Northumberland

Curried Eggs

10 size 3 eggs (hardboiled)
garlic salt (optional)
175ml (6 fl oz) single cream
90ml (6 tbsp) mayonnaise
15ml (1 tbsp) apricot jam
15ml (1 tbsp) mild curry paste
chopped parsley

Halve the hardboiled eggs and lay them face down in a shallow dish. Sprinkle with garlic salt if desired.

Mix the cream, mayonnaise, apricot jam and curry paste together in a bowl. Pour mixture over the eggs. Garnish with chopped parsley.

May be served as a starter or as a light lunch with salad.

Serves 10   *Norfolk*

Baked Stuffed Tomatoes

> 6 large firm tomatoes
> 50g (2oz) cooked ham (chopped)
> 50g (2oz) cheese (grated)
> 50g (2oz) breadcrumbs
> 1 small onion (finely chopped)
> salt and pepper
> mixed herbs

Preheat oven to 200°C / 400°F / Gas mark 6.

Cut the tops off the tomatoes. Carefully scoop out the seeds and pulp with a teaspoon handle.

Mix the tomato pulp with the ham, breadcrumbs, cheese, mixed herbs, onion and seasonings.

Fill tomatoes with the mixture and replace tomato tops.

Bake on a greased tray for 10 minutes at preheated temperature. Then turn down to 180°C / 350°F / Gas mark 4 and continue to cook for 20 minutes.

May also be used as a filling for baked potatoes.

Serves 3–6   *I. Thomas*
                                                  Glamorgan

Ham-stuffed Peaches

411g (16oz) can peach halves
100g (4oz) cooked ham (chopped)
60ml (4 tbsp) mayonnaise
10ml (2 tsp) chopped parsley
salt and pepper
lettuce leaves
parsley sprigs

Drain peaches. Mix ham into mayonnaise with parsley and season to taste.

Spoon mixture into the peach halves, arrange on a bed of lettuce, and garnish with parsley sprigs.

Serves 4

F. Cook
Merseyside

Cheese and Potato Pie

450g (1lb) potatoes
25g (1oz) butter
60ml (4 tbsp) milk
100g (4oz) cheese (grated)

Preheat oven to 190°C / 375°F / Gas mark 5.

Boil potatoes until tender. Mash with butter and then add milk. Add 75g (3oz) cheese.

Place in well-greased 600ml (1pt) pie dish and sprinkle with remaining grated cheese.

Cook until golden brown.

Serves 2–3

J. Dyke
Wiltshire

Crusty Sausages

4 sausages (of choice)
10ml (2 tsp) whole-grain mustard
4 slices crustless white bread
2 tomatoes (skinned and quartered)
50g (2oz) Stilton or cheddar cheese (grated)

Preheat oven 190°C / 375°F / Gas mark 5.

Pre-cook sausages for 15 minutes.

Spread mustard evenly onto one side of bread slices. Roll sausages in bread and place in lightly greased ovenproof dish.

Bake until bread is golden brown and sausages are fully cooked, approximately 15 minutes.

Arrange tomatoes on top of bread, then sprinkle cheese over top.

Return to oven for further 5 minutes until cheese is melted.

Serves 2

E. Guppy
E. Sussex

Cheese Bread and Butter Pudding

6 slices buttered white or brown bread
175g (6oz) grated cheese
300ml (½ pt) milk
2 size 3 eggs

Preheat oven to 190°C / 375°F / Gas mark 5.

Butter a shallow 1.2 litre (2 pt) ovenproof dish.

Line with half the bread, buttered side up, and cover with grated cheese. Put on the remaining bread, buttered side upwards. Beat the milk and the eggs together and pour over the bread.

Leave to rest for 1 hour. Bake for 30 minutes. Eat whilst hot!

Serves 4

B. Harries
Pembroke

Seasoned Pudding

100g (4oz) plain flour
1 size 3 egg
300ml (½pt) milk
salt and pepper
2 medium onions (chopped)
10ml (1 dsp) dried sage
15g (½oz) lard

Preheat oven to 230°C / 450° / Gas mark 8.

Put the flour and seasoning in a basin and make a well in the centre. Add the egg, beating well add the milk gradually until all the flour is incorporated, leaving a smooth mixture.

Boil the chopped onions in a small amount of water until tender. Cool and add to the batter mixture, along with the dried sage.

Put the lard into a 20.5cm (8in.) sandwich tin and place in the oven to heat up.

Remove the tin from the oven, taking care not to burn yourself, and add the batter mixture and cook until brown on top, as for a Yorkshire pudding.

Serve 4

H. Smith
Derbyshire

Salmon Mousse

205g (7½oz) tin red salmon
20ml (4 tsp) gelatine
¼ cup hot water
1 cup mayonnaise
¾ cup plain yoghurt
1 cup finely chopped cucumber

Mash salmon with juices.

Dissolve gelatine in hot water, and add to mayonnaise and yoghurt. Add salmon and cucumber, mix well and leave to set.

Serves 6–8

M. Neville
Kent

Main Meals

Boiled Beef

1350g (3lb) brisket of beef
450g (1lb) onions
450g (1lb) carrots
4 sticks of celery
450g (1lb) swede
450g (1lb) potatoes

Place meat in large pan, ensuring there is enough room for the vegetables.

Cover with cold water and gently bring to the boil. Boil for 10 minutes, spoon off any scum, cover with lid and simmer very gently for 2 hours 15 minutes (35 minutes per pound + 30 minutes).

Add vegetables to pan 1 hour before serving.

Serves 4 with enough left over for the next day.

Dyfed

Beef and Stout Stew with Dumplings

45ml (3 tbsp) sunflower oil
675g (1½lb) stewing steak
1 large onion (sliced)
2 large carrots (sliced)
2 sticks celery (chopped)
45ml (3 tbsp) plain flour
600ml (1 pt) stout
300ml (½pt) beef stock
15ml (1 tbsp) mixed dried herbs

DUMPLINGS
    45ml (3 tbsp) self-raising flour
    50g (2oz) suet
    pinch of salt
    cold water to mix

Preheat oven to 160°C / 325°F / Gas mark 3.

Heat the sunflower oil in a pan. Then fry the meat and onion until browned. Remove meat and onions and place in a medium-sized casserole dish with carrots and celery. Add flour to the fat and cook for a minute. Then add stout and stock and stir until thickened. Pour over meat and vegetables, season and add herbs. Cover and cook gently in oven for 2 hours.

To make dumplings, mix flour, suet and salt with cold water to form a dough. Shape into 4–6 dumplings and place in the casserole.

Cover and cook for a further 20–30 minutes.

Serves 4–6

H. & N. Major
Gloucestershire

Savoury Pudding

    450g (1lb) lean minced beef
    450g (1lb) minced bacon
    225g (8oz) breadcrumbs
    1 size 3 egg (beaten)
    milk to mix
    white pepper and good pinch of nutmeg

Grease a 1.5 litre (2½pt) basin.

Place the minced beef, bacon and breadcrumbs in a large bowl with the egg, pepper and nutmeg. Mix together well, and if rather dry, add a little milk to moisten. Transfer to the greased basin and then place in a steamer and steam for 2 hours.

May be eaten hot or cold.

Serves 6–8

J. Hope
Suffolk

Corned Beef Bake

350g (12oz) potatoes (cubed)
1 225g (8oz) tin corned beef (cubed)
5ml (1 tsp) chilli powder
1 medium onion (chopped)
211g (7½oz) tin tomatoes (chopped)
salt and pepper
butter

Preheat oven to 200°C / 400°F / Gas mark 6.

Boil potatoes until just soft, and drain.

In an 18cm × 28cm (7in. × 11in.) or similar sized ovenproof dish, place the cubes of corned beef, sprinkle the chilli powder over, add the onions and tomatoes.

Place the potato cubes on top, season with salt and pepper and add a few knobs of butter on top. Bake for approximately 20–25 minutes.

Serve with green beans or peas.

Serves 2–3

S. Ling
London

Savoury Ducks

1 cup onions (finely chopped)
1 cup pig's liver (chopped)
1 cup minced beef
1 cup bacon (chopped)
15ml (1 tbsp) sunflower oil
300ml (½pt) beef stock
salt, pepper and mixed spice
1 cup pearl barley
1 cup rice
1 cup sage and onion stuffing mix
600ml (1pt) milk
1 size 3 egg (beaten)
tomatoes and streaky bacon to garnish

Preheat oven to 180°C / 350°F / Gas mark 4.

Fry onions in oil with the liver, mince and bacon until lightly browned. Add beef stock and seasonings.

Meanwhile, cook the pearl barley and rice. Place in a saucepan, cover with cold water, bring to the boil and drain. Then return to saucepan, cover with cold water, bring to the boil, cover pan with lid and simmer for 20 minutes. Remove from heat and drain.

In a large bowl place the rice and pearl barley, the stuffing mix, milk and egg and blend well. Add meat mixture and stir to incorporate.

Divide mixture between two 20.5 × 25.5cm (8in. × 10in.) tins. Level the tops and mark each tin into 8 portions. Top each portion with a slice of tomato and cover that with a slice of streaky bacon, tucking it well in at the sides of each portion.

Leave to stand in refrigerator for 1 hour. Then bake in preheated oven for 30 minutes.

Serve with potatoes, swedes and rich onion gravy.

Freezes well
Each tin serves 8

E. Carr
Yorkshire

Savoury Liver

225g (8oz) liver
1 large cooking apple (peeled and chopped)
1 small onion (chopped)
2.5ml (½ tsp) salt and pepper
2 rashers streaky bacon (chopped)
1 cup water

Preheat oven to 180°C / 350°F / Gas mark 4.

Grease an ovenproof casserole dish.

Cut liver into thin slices. Put into dish, cover with chopped apple, onions and seasoning.

Put chopped bacon on top and add water.

Cover and put in oven for approximately 1–1½ hours. Remove lid for last 20 minutes of cooking time.

Serves 2

V. Shinn
Norfolk

Pan-fried Liver with Bacon

25g (1oz) butter
6 rashers rindless back bacon (chopped)
450g (1lb) lamb's liver (sliced into thin strips)
4 spring onions (cut into 2½cm (1 in.) strips)
½ green pepper (chopped)
100g (4oz) button mushrooms (sliced)
45ml (3 tbsp) dry sherry
15ml (1 tbsp) Worcestershire sauce
15ml (1 tbsp) wholegrain mustard
15ml (1 tbsp) tomato purée
salt and pepper
10ml (2 tsp) chopped fresh sage or 5ml (1 tsp)
    dried sage

Melt the butter in a large pan. Add the bacon and fry for 1 minute. Then add the liver and stir fry for 3 minutes. Add the spring onions, green pepper and mushrooms, stirring constantly for 2 more minutes.

Add the sherry, Worcestershire sauce, mustard, tomato purée, salt and pepper. Cook for a further 3 minutes and then stir in the sage and serve.

Serves 4

M. Charlton
Northumberland

Pork Fillets with Apricots

450g (1lb) pork fillet (cubed)
30ml (2 tbsp) plain flour
50g (2oz) butter
411g (14oz) tin apricots in syrup
30ml (2 tbsp) Worcestershire sauce
30ml (2 tbsp) demerara sugar
10ml (2 tsp) lemon juice
10ml (2 tsp) vinegar
120ml (8 tbsp) water

Preheat oven to 190°C / 375°F / Gas mark 5.

Toss pork in flour. Heat butter in large frying pan, add pork and fry until browned.

Chop apricots. Add any remaining flour to pork and then stir in chopped apricots. Mix 120ml (8 tbsp) apricot juice with Worcestershire sauce, sugar, vinegar, lemon juice and water. Add to pan and bring to boil.

Transfer to casserole dish and cook for 1 hour or until meat is tender.

Serves 4

E. Guppy
E. Sussex

Pork Parcels

100g (4oz) mushrooms (chopped)
1 medium onion (finely chopped)
100g (4oz) bacon (chopped)
450g (1lb) pork fillets
225g (8oz) puff pastry
15ml (1 tbsp) oil
egg yolk

Preheat oven to 220°C / 425°F / Gas mark 7.

Fry mushrooms, onions and bacon in oil until lightly browned.

Remove from pan and put aside. Cut pork fillets into four pieces and brown all over in frying pan.

Roll out puff pastry into four 15cm (6in.) squares. Lay pork fillet in centre of pastry together with mushrooms, onion and bacon mixture. Moisten edges of pastry, bring together and seal parcels.

Place on a baking tray, brush with egg yolk and cook in oven for 30–35 minutes until golden brown. After 10 minutes check that pastry is not browning too quickly; if it is, cover each parcel with foil and continue cooking for remaining time.

Serve with sherry-flavoured gravy, potatoes and a green vegetable.

Serves 4

G. Hoare
Hampshire

Apple and Pork Loaf

450g (1lb) raw minced pork
100g (4oz) rolled oats
450g (1lb) cooking apples
45ml (3 tbsp) water
1 size 2 egg (beaten)
5ml (1 tsp) mixed herbs
2½ml (½ tsp) Tabasco sauce

Preheat oven to 180°C / 350°F / Gas mark 4.

Stew cooking apples in water, sieve when soft.

Place all ingredients in a bowl and blend together.

Press the mixture into a 1kg (2lb) loaf tin and bake for 1¼ hours.

May be served hot with a tomato sauce or cold, garnished with raw apple slices dipped in lemon juice.

Serves 4

F. Cook
Merseyside

Fidget Pie

PASTRY
>   225g (8oz) plain flour
>   50g (2oz) butter or margarine
>   50g (2oz) lard
>   pinch of salt
>   cold water to mix

FILLING
>   225g (8oz) back bacon (roughly chopped)
>   1 medium onion (chopped)
>   225g (8oz) cooking apples (peeled and
>   chopped)
>   15ml (1 tbsp) chopped parsley
>   150ml (¼pt) medium dry cider
>   25g (1oz) plain flour
>   30ml (2 tbsp) brown sugar
>   1 beaten egg to glaze

Preheat oven to 190°C / 375°F / Gas mark 5.

Mix flour and salt in large bowl, rub in fats until mixture resembles breadcrumbs. Using a round-bladed knife to mix, add the cold water to form a stiff dough. Turn dough onto floured surface, and knead lightly. Wrap in foil, and chill for 30 minutes.

Put bacon, onion and apples into a 600ml (1pt) pie dish. Add parsley and season to taste.

Blend flour and sugar with the cider a little at a time, and then pour into pie dish.

Roll out pastry. Moisten rim of dish, then place a thin strip of pastry all the way around the rim, press down lightly. Moisten the strip, place pastry lid on top, press to seal and flute the edge, or place a pie funnel inside and cut a cross in the top to let the steam out.

Brush the pastry with the egg and bake in oven for 45 minutes, or until pastry is golden brown, and filling is cooked.

Serves 4

Margaret Bywater
Yorkshire

Lamb Hotpot

900g (2lb) lamb neck fillet
2 lamb's kidneys
900g (2lb) potatoes (sliced)
50g (2oz) mushrooms (sliced)
salt and pepper
150ml (¼pt) stock
butter

Preheat oven to 180°C / 350°F / Gas mark 4.

Trim the meat of any fat or bone and cut into bite-size cubes. Halve the kidneys, remove the core and skin and cut into slices. Layer the potatoes and mushrooms with the lamb and kidneys in a large 2½ litre (4–5pt) casserole, season well and finish up with a layer of potatoes.

Pour over the stock and dot the potatoes with a little butter. Cover the casserole dish with a lid and bake in the oven for 1½ hours. Remove the lid and continue baking for a further 30 minutes to brown the potatoes.

Serves 4

L. Griffiths
Dyfed, Wales

Lamb – Japanese style

15ml (1 tbsp) oil
2 lamb fillets
1 bunch spring onions, trimmed and cut into
    2½cm (1in.) lengths
15ml (1 tbsp) soy sauce
75g (3oz) butter
juice of ½ lemon
½ small red pepper (diced)
100g (4oz) broccoli florets
½ small yellow pepper (diced)
100g (4oz) mushrooms (sliced)
15ml (1 tbsp) roasted peanuts
5ml (1 tsp) grated fresh root ginger
15ml (1 tbsp) clear honey
10ml (2 tsp) fresh coriander (chopped)
5ml (1 tsp) sesame seeds

Heat the oil in a large frying pan. Add the lamb fillets and cook for
5 minutes, turning until brown on all sides.

Add the spring onions and soy sauce and cook for a further 6 minutes,
or until the lamb fillets are tender and just cooked.

Meanwhile, heat the butter in a pan until it just starts to bubble.
Then add the lemon juice, peppers, broccoli florets, mushrooms,
peanuts, ginger, honey and coriander, mixing well. Cook for
30 seconds and then remove from the heat.

Cut the lamb fillets diagonally with a sharp knife into thick slices and
put on a serving plate. Spoon over the pepper mixture, sprinkle with
sesame seeds and serve immediately.

Serves 4

M. Charlton
Northumberland

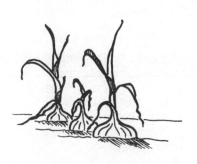

Herbed Lamb Cutlets

8 × 100g (4oz) lamb cutlets
2 cloves garlic (peeled and halved)
15ml (1 tbsp) olive oil
15ml (1 tbsp) mixed fresh thyme, marjoram
   and rosemary, or
5ml (1 tsp) dried mixed herbs
freshly ground black pepper
fresh mint and lemon wedges to garnish

Lay cutlets in shallow dish, and rub all over with the cut side of the garlic cloves.

Brush both sides of the cutlets with oil, coat with the herbs and season with pepper.

Place the cutlets on grill rack, then grill for 4 minutes on each side. Serve with new potatoes and green salad.

Serves 4

Powys

Lamb and Rice Bake

STUFFING
15ml (1 tbsp) chopped thyme and parsley
2 cups white breadcrumbs
grated rind and juice of 1 lemon
100g (4oz) sausage meat
15ml (1 tbsp) brown sugar
15ml (1 tbsp) Worcestershire sauce
1 size 3 egg (beaten)

shoulder of lamb (boned and trimmed of any
    fat)
salt and pepper
30ml (2 tbsp) sunflower oil
600ml (1pt) chicken stock
175g (6oz) long grained rice
parsley and marjoram sprigs

Preheat oven to 180°C / 350°F / Gas mark 4.

Prepare stuffing by mixing all the ingredients together in a bowl.

Pack firmly into the shoulder of lamb. Sew edges together with thread or use skewers, if preferred, to prevent stuffing falling out.

In a large pan, heat the sunflower oil and brown the meat all over.

Pour stock into a large casserole dish and add rice. Place lamb in centre, pushing it well into the rice. Add the herbs, cover and bake for 2 hours.

Serves 6

H. & N. Major
Gloucestershire

Chicken Tarragon

40g (1½oz) butter
30ml (2 tbsp) sunflower oil
4 chicken breasts
1 large onion (chopped finely)
175ml (6fl oz) orange juice
150ml (¼pt) water
1 chicken stock cube
4 sprigs fresh tarragon or 15ml (1 tbsp) dried
    tarragon
150ml (5fl oz) soured cream
10ml (2 tsp) cornflour

Preheat oven to 180°C / 350°F / Gas mark 4.

Melt butter and oil in a medium-sized casserole dish and brown chicken breasts on both sides. Remove skin if desired. Lift out chicken and set to one side.

Gently fry the onion in the remaining fat. Stir in the orange juice, water, stock cube and tarragon, and gently bring to the boil.

Return the chicken to the casserole dish and bake in oven for 1 hour, basting occasionally.

Remove the chicken and place on a serving dish. Cover with foil and keep warm whilst making sauce.

Blend the cornflour with 30ml (2 tbsp) of water and add to juices in the casserole. Carefully bring to the boil and stir continuously for 2–3 minutes. Cool slightly before adding the soured cream.

Pour over the chicken and garnish with fresh tarragon or parsley.

Serves 4

F. Cook
Merseyside

Chicken Fricassee

2 chicken breasts
1 bay leaf
2½ml (½ tsp) lemon juice
150ml (¼pt) chicken stock
1 small onion (finely chopped)
50g (2oz) mushrooms (sliced)
25g (1oz) flour
25g (1oz) butter
150ml (¼pt) milk
salt and black pepper

Remove the skin from the chicken and place in a medium-sized saucepan. Add the chopped onion, mushroom slices, bay leaf, salt and pepper, lemon juice and stock. Bring to the boil and then simmer for 20 minutes until the chicken is tender. Remove the chicken and vegetables and strain the liquid into a measuring jug. Cut the chicken up into bite-size pieces.

Add the milk to the stock to make up to 300ml (½pt). Make a white sauce by melting the butter in a small saucepan. Add the flour, beating all the time, and cook for one minute, then slowly add the milk/stock. Bring to the boil and cook gently for 2 minutes, stirring all the time. Return the chicken, mushrooms and onions into the white sauce and heat thoroughly.

Serve with boiled rice.

Serves 2–3

E. Holmes
Cambridgeshire

Bengal Chicken

15ml (1 tbsp) oil
2 onions (peeled and chopped)
1 clove garlic
5ml (1 tsp) curry powder (mild or strong)
5ml (1 tsp) dry mustard powder
4 chicken breasts
300ml (½pt) chicken stock (300g)
30ml (2 tbsp) tomato purée
salt and pepper

Preheat oven to 180°C / 350°F / Gas mark 4.

Heat oil and sauté the onions and garlic until soft.

Mix together curry powder and mustard and rub into chicken breasts. Add to pan and sauté until golden.

Mix together stock and tomato purée and pour over chicken. Season to taste.

Place in an ovenproof casserole dish and cook for 1 hour.

Serves 4

Berkshire

Devilled Chicken Legs

4 chicken leg portions
225g (½lb) mushrooms
15ml (1 tbsp) sunflower oil

SAUCE
45ml (3 tbsp) sunflower oil
15ml (1 tbsp) Worcestershire sauce
15ml (1 tbsp) tarragon vinegar
1 finely chopped onion
salt and black pepper

Mix all sauce ingredients together.

Score the chicken portions with a sharp knife and place in a grill pan. Brush chicken with the sauce, and grill for approximately 10 minutes on each side, basting often.

Five minutes before the end of cooking, brush the mushrooms with sunflower oil, place them in the grill pan and grill for 5 minutes until cooked.

Pour any devilled sauce left in the grill pan over the chicken. Serve with noodles and a green salad.

Serves 4

E. Drowne
Devon

Chicken and Lemon Casserole

4 chicken joints
salt and pepper
15ml (1 tbsp) chopped lemon thyme
grated rind and juice of 1 lemon
600ml (1pt) chicken stock

15ml (1 tbsp) cornflour
150ml (¼pt) crème fraiche
lemon wedges for garnish

Preheat oven to 180°C / 350°F / Gas mark 4.

Place chicken joints in a medium-sized casserole, and rub in salt, pepper and lemon thyme. Sprinkle over the rind and juice of the lemon and add the chicken stock. Cook in the oven for 1 hour.

Drain the liquid from the casserole into a saucepan. Return the chicken to the oven to keep warm.

Mix cornflour with cold water to a smooth paste, and stir into the stock.

Gradually bring liquid to the boil, stirring all the time. Reduce heat and add crème fraiche. Season to taste.

Pour sauce over chicken joints and serve immediately, garnished with lemon.

Serves 4   *Cambridgeshire*

Glazed Lemon Chicken

1 clove garlic (crushed)
15ml (1 tbsp) fresh rosemary (finely chopped)
salt and black pepper
grated rind and juice of 2 lemons
15ml (1 tbsp) olive oil
5ml (1 tsp) sugar
30ml (2 rounded tbsp) Seville fine cut
    marmalade
bunch of spring onions or 6 shallots (chopped)
4 chicken joints
10ml (2 tsp) cornflour

Make a marinade of crushed garlic, rosemary, seasoning, grated rind and juice of the lemons, olive oil, sugar and marmalade in a shallow dish and blend with a fork. Add the chicken and turn to coat evenly. Cover with cling film and refrigerate for at least 1 hour or overnight. Turn the chicken in the marinade from time to time.

Remove the chicken from the marinade and pat dry with kitchen paper. Peel and chop the onions. Heat some olive oil in a large pan and fry the chicken and onions until golden brown all over. Add marinade to the pan and bring to a simmer. Cover and cook gently for approximately 15–20 minutes.

If desired, the sauce may be thickened with 10ml (1 dsp) cornflour mixed with a little water.

Serve with mashed potatoes and a green vegetable.

Serves 4

B. Rudman
East Sussex

Indian Curry

CURRY POWDER
     25g (1oz) cayenne
     50g (2oz) ground ginger
     225g (8oz) coriander seed
     50g (2oz) mustard powder
     100g (4oz) fenugreek seed
     100g (4oz) cinnamon
     225g (8oz) turmeric

     3 large onions (finely chopped)
     45ml (3 tbsp) sunflower oil
     30ml (2 tbsp) curry powder
     1 cup stock
     15ml (1 tbsp) chutney
     15ml (1 tbsp) apricot jam
     30ml (2 tbsp) vinegar
     5ml (1 tsp) sugar
     30ml (2 tbsp) milk
     450g (1lb) cooked meat or chicken (cubed)

To make curry powder, grind ingredients using pestle and mortar. Bottle and seal well.

Brown onions in the sunflower oil in a large saucepan and add the curry powder. Cook for 10 minutes, stirring constantly so that the mixture does not burn. This cooking prevents the raw taste sometimes found in curries. Gently add the stock and cook for 5 minutes, stirring all the time. Add the chutney, jam, vinegar and sugar and cook for a further 10 minutes.

Add meat or chicken. Bring back to the boil and simmer for at least 30 minutes until completely cooked through.

Note that the longer the curry simmers, the better it becomes. Take care not to allow it to burn.

Just before serving add the milk, which softens the taste.

ACCOMPANIMENTS
boiled rice
chilli butter (creamed butter with finely
    chopped green chillies)
chutney
Bombay duck
cucumber in coconut milk
raisins
grated coconut
sliced banana
chapatties
strawberry jam
nuts
sliced ripe tomatoes

Serves 3–4

Mrs Squirrel
Suffolk

Rabbit Pie

5ml (1 tsp) oil
5ml (1 tsp) butter
1 rabbit or 450g (1 lb) diced rabbit
15ml (1 tbsp) seasoned flour
225g (8oz) streaky bacon
1 large onion (sliced)
2 carrots (sliced)
½ swede (cubed)
600ml (1 pt) beef stock
fresh marjoram
225g (8oz) sliced mushrooms
4 thick slices bread
parsley to garnish

Preheat oven to 180°C / 350°F / Gas mark 4.

Melt butter and oil in a large frying pan. Coat rabbit in seasoned flour and fry with bacon until brown. Transfer to medium-sized casserole.

Add onions, carrots and swede to the pan, stir for a few minutes. Add stock and bring to boil, add marjoram and season. Pour over meat.

Cover and cook for 1½ hours. Remove from oven and add mushrooms. Cut bread into triangles, dip into gravy and place, gravy side up, overlapping neatly on top of the meat. Return to the oven, and cook uncovered for a further 30 minutes until bread is crisp.

Sprinkle with chopped parsley.

Serves 4–6

H. & N. Major
Gloucestershire

Poached Trout in Cider

2 fresh cleaned trout, approximately 350g
(12oz) each
600ml (1 pt) dry cider

STUFFING
½ medium onion (finely chopped)
chopped parsley
juice of half a lemon
25g (1oz) fresh white breadcrumbs

SAUCE
1 egg yolk
10ml (1 dsp) plain yoghurt
10ml (1 dsp) double cream
season to taste

Preheat oven to 200°C / 400°F / Gas mark 6.

Mix the stuffing ingredients together, seasoning lightly with salt and pepper to taste. Fill the cavity of each trout. Sew up the trout with thread to stop the stuffing falling out.

Place the fish side by side in a ovenproof dish and pour over the cider.

Cover the dish with either a lid or foil and place in a large roasting tin containing water that has already come to simmer in the oven and that comes halfway up the side of the ovenproof dish. Poach for about 30 minutes.

Remove fish and place in a covered dish to keep warm.

To make sauce, drain off cider into a medium saucepan and at high heat reduce to half the quantity. Take off the heat and stir in the egg yolk, yoghurt, cream and seasoning to taste. Stir over a low heat to thicken slightly, but do not boil.

Remove the thread from the fish and pour sauce over. Serve with plain boiled potatoes and green vegetables.

Serves 2

D. Quaife
E. Sussex

Puddings

Trifle

450g (1lb) loaf-tin size Madeira cake
12 ratafia biscuits (crumbled)
150ml (¼pt) sweet sherry
75ml (2½ fl oz) orange juice
60ml (4 tbsp) strawberry jam or conserve
(homemade preferably)
45ml (3 tbsp) cornflour
60ml (4 tbsp) caster sugar
300ml (½pt) milk
300ml (½pt) single cream
3 size 3 eggs
5ml (1 tsp) vanilla essence
5ml (1 tbsp) butter
300ml (½pt) whipping cream
toasted flaked almonds

Mix together the sherry and the orange juice in a small bowl.

In the base of a large trifle bowl place the Madeira cake in slices. Sprinkle with crumbled ratafia biscuits. Pour over sherry and orange juice. Spread the jam evenly over the top. Leave in refrigerator for cake to soak up the liquids.

Meanwhile make the custard. Blend together the cornflour and the caster sugar with a little of the milk, and mix to a smooth paste. Put water in the base of a double saucepan on to simmer, in the top of the double saucepan heat the cream and the remaining milk. When the milk comes to the boil pour it over the cornflour paste and stir well. Return to the pan and bring to the boil, stirring all the while as the custard thickens. Boil for 2 minutes. Remove from heat.

Separate eggs, place the egg yolks in the custard and beat in. Return custard to heat and cook for 10 minutes, stirring occasionally. Strain sauce into a bowl and stir in vanilla essence and butter. Leave to cool. Sprinkle a tiny amount of sugar on top of custard to prevent a skin forming.

Beat the egg whites until stiff but not too dry and fold into the cooled custard. Pour over the soaked cake and jam, and level the top with a spoon. Refrigerate until custard has set awhile. Whip the whipping cream until thick, then pour over top of custard.

Decorate with toasted flaked almonds.

Serves 8

D. Froome
Berkshire

Pineapple Pudding

425g tin crushed pineapple
12½g (½oz) sachet of gelatine
4 size 2 eggs (separated)
100g (4oz) granulated sugar
juice of ½ lemon
250ml (½pt) double cream
small tin pineapple chunks (for decoration)

Drain crushed pineapple and reserve 225ml (8fl oz) juice.

Prepare gelatine. Sprinkle gelatine on 60ml (4 tbsp) pineapple juice, and dissolve in a bowl over hot water. When mixture goes clear, the gelatine is dissolved. Cool slightly.

Whisk egg yolks and sugar until thick and creamy. Add remaining pineapple juice, lemon juice and the gelatine mixture, and stir thoroughly. Leave to thicken for 8–10 minutes.

Whip the cream to the floppy stage. Whip the egg whites until stiff but not dry. When the gelatine mixture has thickened, fold in the cream, and then fold in the egg whites.

Add the crushed pineapple carefully. Pour into a large bowl and refrigerate until set.

Decorate with pineapple chunks.

Serves 6

D. Perron
Hampshire

Lemon Soufflé

12½g (½oz) sachet of gelatine
60ml (4 tbsp) very hot water
3 size 2 eggs (separated)
200g (7oz) caster sugar
grated rind and juice of 2 large lemons
300ml (½pt) double cream

Prepare a 900ml (1½pt) soufflé dish by securing a baking parchment collar above the top of the dish. Lightly grease the insides.

To prepare gelatine, put water in basin, sprinkle in the gelatine and dissolve over a saucepan of hot water. Cool slightly.

Whisk egg yolks and sugar together until thick and creamy. Add lemon rind and juice, pour in gelatine mixture, mix in gently but thoroughly.

Whisk cream to the floppy stage, whisk the egg whites until stiff. Fold cream into the yolk and gelatine mixture, then fold in the egg whites. Pour into the prepared dish. Leave to set in the refrigerator.

When ready, carefully remove paper collar. Decorate if desired with cream rosettes.

Serves 6

D. Perrott
Hampshire

$\mathcal{D}$amson $\mathcal{C}$ream

450g (1lb) damsons
100ml (4fl oz) water
1 packet black cherry jelly
150g (5oz) black cherry yoghurt
150g (5oz) whipping cream (whipped)

Stew damsons in water. Drain, reserving juice, and press through sieve to purée. Add enough juice to make purée up to 300ml (½pt).

Melt jelly in a small amount of damson juice, add to fruit purée and stir well. Stir in cherry yoghurt and then fold in whipping cream.

Place in serving dish and refrigerate to set.

Can be made with any fruit/yoghurt/jelly combinations.

Serves 4

E. Southam
Cambridgeshire

Nanna McNeill's Creams

1 packet raspberry jelly
600ml (1pt) custard
cherries and cream to decorate

Mix jelly with 450ml (¾pt) boiled water. When jelly has dissolved, but is still warm, stir in the custard. Mix thoroughly.

Pour into individual glasses. When cool, decorate with cherries and whipped cream.

Serves 6   *Suffolk*

Special Apple Dessert Cake

150g (5oz) butter
2 size 2 eggs
225g (8oz) caster sugar
5ml (1 tsp) almond essence
225g (8oz) self-raising flour
7½ml (1½ level tsp) baking powder
675g (1½lb) cooking apples (peeled, cored
    and sliced)
icing sugar

Preheat oven to 160°C / 325°F / Gas mark 3.

Grease and line a 20.5cm (8in.) loose-bottomed cake tin.

Melt the butter in a pan over medium heat until just runny, and pour into a large bowl. Add the eggs, sugar and almond essence and beat until well mixed. Fold in the flour and baking powder.

Spread two-thirds of the mixture into the prepared cake tin. Place the sliced apples on top and then spread the remaining mixture over the apples.

Bake for 1½ hrs until the apple is tender when a skewer is inserted in the cake. Carefully remove from the cake tin and dredge with the icing sugar when cooled.

May be served warm or cold with thick cream.

Serves 8–10

L. Griffiths
Dyfed

✐Mandarin Cheesecake

BASE
225g (8oz) digestive biscuits (crushed)
100g (4oz) butter

FILLING
225g (8oz) cream cheese
75g (3oz) caster sugar
1 300g (11oz) can mandarin oranges
1 packet orange jelly
150g (5oz) carton natural yoghurt
150ml (¼pt) double cream
2 size 3 egg whites

Over a gentle heat, melt butter, remove from heat and add biscuits. Put into base of 20.5cm (8in.) flan tin, firm and place in fridge to set.

Beat cream cheese and sugar together.

Drain and chop the mandarins (reserving a few for decoration). Reserve juice.

Heat 45ml (3 tbsp) of juice with the cut-up jelly over a gentle heat until dissolved. Cool, and then stir the jelly into the cream cheese mixture, along with the yoghurt. Fold in the lightly whipped double cream.

Whisk egg whites until stiff, and fold into mixture. Spoon over biscuit base, and refrigerate until set.

Decorate with reserved mandarins.

Serves 8

Powys

Parson's Folly

BASE
>225g (8oz) ginger biscuits (crushed)
>175g (6oz) butter
>pinch of salt
>15ml (1 tbsp) water

FILLING
>Grated rind and juice of 1 lemon
>100g (4oz) hazelnuts (chopped)
>2 size 3 eggs
>50g (2oz) caster sugar
>150ml (5fl oz) double cream

TOPPING
>450g (1lb) fresh or frozen raspberries
>15ml (1 tbsp) arrowroot
>1 measure gin

Preheat oven to 180°C / 350°F / Gas mark 4.

Melt butter, and blend in biscuit crumbs, salt and water. Put in base of 20.5cm (8in.) flan tin. Firm down and refrigerate.

Meanwhile make filling. Place lemon rind, juice, cream, chopped hazelnuts and eggs in a bowl, and whisk for 2 minutes. Pour over biscuit base and bake for 20 minutes.

Remove from oven and cool.

Drain juice from raspberries (if frozen).

To make topping, add enough water to raspberry juice to make 150ml (5fl oz). Place in a small saucepan and blend with arrowroot and bring to the boil for 1 minute. Remove from heat and stir in gin. Place raspberries over cream filling, then spoon sauce topping over.

Refrigerate before serving.

Serves 6–8

M Horn
Norfolk

Apricot Cheesecake

225g (8oz) ginger biscuits (crushed)
75g (3oz) butter (melted)
1 orange jelly
411g (14½oz) tin apricots
225g (8oz) cottage cheese
25g (1oz) caster sugar
150ml (¼pt) double cream (lightly whipped)

Prepare a loose-bottomed 18cm (7in.) flan tin.

Mix crushed biscuits and butter and press into base of flan tin. Cover and put in fridge to set.

Drain apricots, reserving 45ml (3 tbsp) of the juice. Dissolve jelly and the apricot juice in a bowl over hot water. Push cottage cheese and apricots through a sieve and mix with caster sugar and jelly mixture when cool. Allow to set slightly, then fold in cream.

Put mixture into flan dish, and put into fridge to set approximately 30 minutes. Decorate as desired.

Serves 6–8

H. Allen
Lincolnshire

Pineapple Meringue

175g (6oz) crushed digestive biscuits
75g (3oz) butter
30ml (2 tbsp) custard powder
30ml (2 tbsp) caster sugar
large can crushed pineapple
2 size 3 eggs (separated)
75g (3oz) caster sugar

Preheat oven to 100°C / 200°F / Gas mark ¼.

Melt butter and add to crushed biscuits, press into a 20.5cm (8in.) flan dish, and put into fridge to set.

Mix custard powder and caster sugar with juice from pineapple, and cook over a low heat until thickened.

Remove from heat and cool. Then add egg yolks and pineapple, and put into flan dish.

Make meringue by whisking egg whites until stiff and folding in caster sugar. Spread over mixture and bake in oven until meringue is golden brown.

Serves 6–8

Cambridgeshire

Pear and Apple Crumble

450g (1lb) cooking apples (peeled, cored and
  chopped)
450g (1lb) pears (peeled, cored and
  chopped)
30–45ml (2–3 tbsp) granulated sugar

TOPPING
175g (6oz) plain flour
75g (3oz) butter
40g (1½oz) caster sugar

Preheat oven to 200°C / 400°F / Gas mark 6.

Put apples in saucepan and add 125ml (4 fl oz) water. Bring to boil, cover and simmer over gentle heat for 5 minutes.

Drain apples, add to pears and sugar and put into a 1.2 litre (2pt) ovenproof dish.

Sift flour into a mixing bowl and rub in butter until mixture resembles fine breadcrumbs. Stir in sugar.

Spoon crumble mix over fruit. Bake for 35–40 minutes, or until crumble is golden brown.

Serves 6–8

E. Drowne
Devon

Apricot and Marzipan Slice

225g (8oz) packet puff pastry
75g (3oz) marzipan
400g (14oz) tin apricot halves in syrup
flaked almonds

Preheat oven to 200°C / 400°F / Gas mark 7.

Roll out puff pastry and put in base of a 23cm × 33cm (9in. × 13in.) Swiss roll tin.

Roll out marzipan and place on top of pastry, leaving a 2.5cm (1in.) border.

Strain apricots and place face down on top of marzipan. Brush all over with apricot syrup, sprinkle with flaked almonds and bake in the oven for 20 minutes.

Serves 4–6

J. Hicks
Devon

Apple Flan

225g (8oz) frozen puff pastry
3 medium-sized cooking apples (peeled, cored and grated)
30ml (2 tbsp) golden syrup
grated rind of 1 lemon
icing sugar

Preheat oven to 180°C / 350°F / Gas mark 4.

Roll out pastry and line a Swiss roll tin 23cm × 33cm (9in. × 13in.).

Place half of the grated apples into the pastry case with the grated lemon rind and evenly drizzle over the golden syrup. Top with the remaining grated apples.

Bake for 20–25 minutes, watching carefully to ensure that the edges do not burn.

When slightly cool, sprinkle with icing sugar. Serve warm or cold with cream.

Serves 6–8

Mrs Salisbury
Clwyd

Raisin Tart

PASTRY
> 100g (4oz) plain flour
> 25g (1oz) butter or margarine
> 25g (1oz) lard
> pinch of salt
> cold water to mix

FILLING
> 225g (8oz) raisins
> 50g (2oz) butter
> 50g (2oz) soft light brown sugar
> 1 size 2 egg (beaten)

Preheat oven to 200°C / 400°F / Gas mark 6.

Mix flour and salt in large bowl, and rub in fat until mixture resembles breadcrumbs. Using a round-bladed knife to mix, add the cold water to form a stiff dough.

Turn dough onto floured surface and knead lightly. Roll out and line an 18cm (7in.) flan dish.

Place raisins in pastry case. Cream butter and sugar, and beat in egg. Then pour mixture over raisins.

Bake for 20 minutes. Serve warm.

Serves 8

A. Hambrook
E. Sussex

Gainsborough Tart

PASTRY
     175g (6oz) plain flour
     40g (1½oz) butter or hard margarine
     40g (1½oz) lard
     pinch of salt
     cold water to mix

FILLING
     25g (1oz) butter
     50g (2oz) caster sugar
     1 size 3 egg (beaten)
     100g (4oz) desiccated coconut
     2½ml (½ tsp) baking powder
     30ml (2 tbsp) raspberry jam

Preheat oven to 180°C / 350°F / Gas mark 4.

Mix flour and salt in large bowl, and then rub in fats until mixture resembles breadcrumbs.

Using a round-bladed knife to mix, add cold water to form a stiff dough. Turn out onto floured surface and knead lightly. Roll out and line a 20.5cm (8in.) flan dish.

Cream butter and sugar and beat in egg. Fold in desiccated coconut with baking powder.

Spread jam over pastry base, and then evenly pour filling over. Bake for 30 minutes.

Serve cold or just warm.

Serves 8   *Cornwall*

Lemon Surprise Pudding

50g (2oz) butter
150g (5oz) caster sugar
30ml (2 tbsp) hot water
50g (2oz) self-raising flour
juice and rind of 1 lemon
2 size 3 eggs (separated)
150ml (¼pt) milk

Preheat oven to 180°C / 350°F / Gas mark 4.

Lightly butter a 1.2 litre (2pt) pie dish.

Beat sugar and butter well, adding hot water to slacken.

Add flour, juice and rind of lemon, egg yolks and milk and beat well. Whisk egg whites until stiff and fold into the mixture.

Place in pie dish and bake for 45 minutes.

Serves 4

South Yorkshire

Mandarin Pudding (for microwave)

3 slices white bread (crusts removed)
50g (2oz) brown sugar
50g (2oz) butter
298g (8oz) tin mandarin oranges (drained)
25g (1oz) sultanas
2 size 3 eggs
300ml (½pt) milk

Put the butter in a deep round ovenproof dish, microwave for 1–1½ minutes to melt.

Cut bread into triangles and place in dish, turning pieces over so they are all well coated in the butter. Arrange them in the dish, and sprinkle with all but 30ml (2 tbsp) of the brown sugar. Arrange the mandarins and sultanas on top. Beat the eggs and milk together and pour over the fruit.

Microwave on a medium setting for 10–12 minutes, until the custard has set at the edges but is still slightly damp in the centre.

Leave to stand for 5 minutes. Sprinkle with the remaining sugar and place under a preheated grill to brown the top.

Serve hot.

Serves 4   *Powys*

Queen of Bramble Pudding

30ml (2 tbsp) water
60ml (4 tbsp) sugar
450g (1lb) blackberries
5ml (1 tsp) cornflour
100g (4oz) white bread, cut into 1cm (½in.) cubes
600ml (1 pt) milk
50g (2oz) unsalted butter
3 size 3 eggs (separated)
grated rind of small lemon
150g (5oz) caster sugar

Preheat oven to 160°C / 325°F / Gas mark 3.

Lightly butter a 1.2 litre (2pt) pie dish.

Stew blackberries lightly in the sugar and water. Then purée the mixture through a sieve **or** pour off the liquid from the blackberries and set them aside. Thicken the purée or liquid with cornflour and reheat.

Place bread in pie dish. Heat milk and unsalted butter until just warm. Add beaten egg yolks, lemon rind and 75g (3oz) sugar. Pour this custard over bread.

Bake pudding for 20 minutes until custard is set. Remove from oven.

Spread purée or whole blackberries over pudding.

Whisk egg whites until stiff, and fold in remaining sugar. Pile this meringue over the baked pudding. Return to oven to brown meringue lightly.

Serve with cream and the bramble sauce if desired.

Serves 4

J. Goddard
Hampshire

Eve's Orange Pudding

900g (2lb) cooking apples
1 orange
100g (4oz) granulated sugar
100g (4oz) soft margarine
100g (4oz) caster sugar
175g (6oz) self-raising flour
pinch of salt
2 size 3 eggs
30ml (2 tbsp) milk
75g (3oz) icing sugar

Preheat oven to 180°C / 350°F / Gas mark 4.

Peel and core apples, and cut into thin slices. Grate orange rind and mix into granulated sugar. Lightly butter a 1.2 litre (2pt) ovenproof dish. Layer apple slices and orange-flavoured sugar, finishing with a layer of apple.

Put all the remaining ingredients except the icing sugar into a bowl and beat well for approximately 2 minutes. Cover the apples with the sponge mixture.

Bake for 1 hour until sponge is well risen and golden brown, and springs back when pressed with a finger.

When pudding is removed from the oven, squeeze juice from orange, mix into icing sugar and pour over the top. Serve hot.

Serves 6

E. Guppy
E. Sussex

Golden Pudding

50g (2oz) soft margarine or butter
50g (2oz) brown sugar
½ teacup golden syrup
5ml (1 tsp) vanilla essence
2 size 3 eggs
200g (7oz) self-raising flour
2½ml (½ tsp) mixed spice
½ teacup of milk
45ml (3 tbsp) sugar
15ml (1 tbsp) grated orange rind

SAUCE

15ml (1 tbsp) cornflour
½ teacup sugar
150ml (½pt) water
salt
1 lemon (grated rind and juice)
25g (1oz) butter or margarine

Preheat oven to 160°C / 325°F / Gas mark 3.

Grease and line a 20.5cm (8in.) ovenproof dish.

Cream margarine and sugar together until light and fluffy. Add vanilla essence beating well.

Add eggs, beating well. Fold in flour and spice and add milk.

Put in prepared dish and sprinkle with 45ml (3 tbsp) sugar mixed with grated orange rind.

Bake for 1 hour or until an inserted skewer in the centre comes out clean.

For the lemon sauce mix cornflour with sugar in a small saucepan. Stir in boiling water, add a pinch of salt and bring to the boil. Simmer, stirring occasionally, for 15 minutes. Stir in grated lemon rind, juice and butter or margarine.

Cut into squares and serve with lemon sauce.

Serves 8

M. Adams
Builth Wells

Banana Pudding

3 cups breadcrumbs
1 cup sultanas
½ cup white sugar
50g (2oz) glacé cherries (halved)
3 ripe bananas (mashed)
25g (1oz) butter (melted)
5ml (1 tsp) bicarbonate of soda
pinch of salt

Lightly butter a 1.2 litre (2pt) pudding basin.

Mix thoroughly all the ingredients. Place in the pudding basin and cover with a layer of greaseproof paper and foil tied with string.

Steam for 1½–2 hours. (Don't forget to check water levels during cooking.)

Serves 6

Powys

Apple Gob

225g (8oz) self-raising flour
100g (4oz) shredded suet
pinch of salt
cold water to mix
900g (2lb) cooking apples (peeled, cored and sliced)
50g (2oz) sugar

Mix flour, salt, and suet together in a bowl and add enough water to make a soft dough. Turn dough onto a floured surface and knead lightly.

Roll out and line a 1.2 litre (2pt) pudding basin, reserving enough rolled out dough for the top.

Place the thinly sliced apples into the pastry lined bowl, dredging with sugar, add a little water and seal with top layer of pastry.

Cover pudding with a layer of greaseproof paper and a layer of foil tied with string, place in a steamer and steam for 2 hours.

Serve with hot custard.

Serves 4–6

A. Davis
Dyfed

ℱairy 𝒟atε 𝒫udding

225g (8oz) plain flour
1ml (¼ tsp) salt
2½ml (½ tsp) bicarbonate of soda
1ml (¼ tsp) mixed spice
1ml (¼ tsp) nutmeg
1ml (¾ tsp) cinnamon
100g (4oz) butter or soft margarine
225g (8oz) dates (stoned and chopped)
150ml (¼pt) milk
1 size 3 egg (beaten)
175g (6oz) golden syrup

Sieve together the plain flour, salt, bicarbonate of soda and spices. Rub in the butter until the mixture resembles fine breadcrumbs. Stir in the dates, milk, beaten egg and golden syrup. Mix thoroughly to blend all the ingredients together. Turn into a lightly greased 900ml (1½ pt) pudding basin. Cover with greaseproof paper and foil and tie securely with string. Place in a steamer and cook for 2 hours.

Serves 6

M. Bills
Bedfordshire

Cakes

By Request?

A recipe is needed for your forthcoming book,
I've searched my brain every cranny and nook,
For over thirty years I made bread of renown,
Cottage loaves crusty but moist, a rich golden brown.

But I'm sorry to say all done by rote,
The method by now I'm unable to quote.
How about a jam sandwich made by my Gran?
When she made bread she always added some bran.

Method:
Take three medium-sized eggs, of course range free,
No batteries those days, proper eggs for tea.
Weight of two eggs in sugar of the caster kind,
Weight of one egg in flour, seems mean to my mind.

Line the bottom of tins with paper well greased,
The easiest part of the job has now ceased.
Separate yolks and whites with the utmost care,
You probably now have not much time to spare.

Add a pinch of salt to the whites you'll be beating.
You can do all this whilst the oven is heating.
Then whisk to soft points and your arm goes numb,
Oh! of course you're electric . . . my brain goes dumb!

The beaten yolk fold in with a spoon of wood,
Gran says 'it be better' – can't see why it should.
Next beat in sugar, don't give it time to settle,
Fold in sifted flour, use large spoon of metal.

This wee secret Gran kept for many a year,
(you're lucky to get this information, dear)
A kettle of boiling water at the ready,
A large table spoon in the hand steady.

A spoon of hot water tipped into the lot,
Quickly tin it . . . into the oven hot.
When all seems firm, twelve minutes more or less,
Turn onto sugared paper, should be no mess.

As it starts to cool remove the greased paper,
Then do the jammy bit, you know the caper.
If this does not work don't curse me with rage,
The memory is not what it was. At my age !!

P.S. If in need slip quickly to the local shop
and make 'VIOTA'* buns with a cherry on the top.

* Once a popular make of cake mix.

Mrs D. Bacon
Norfolk

✨ll-in Fruit Cake

225g (8oz) self-raising flour
100g (4oz) caster sugar
350g (12oz) mixed fruit
50g (2oz) glacé cherries
2 size 3 eggs
1½ml (¼ tsp) mixed spice
150ml (6fl oz) milk
2½ml (½ tsp) vanilla essence
15ml (1 tbsp) sherry
100g (4oz) melted margarine or butter

Preheat oven to 140°C / 275°F / Gas mark 1.

Grease and line a 20.5cm (8in.) cake tin.

Place all ingredients in a large bowl and beat well together. Spoon
into prepared tin and bake for approximately 2 hours. When cake is
cooked an inserted skewer should come out clean.

Serves 12

Farmhouse Soda Cake

225g (8oz) plain flour
pinch of salt
2½ml (½ tsp) bicarbonate of soda
2½ml (½ tsp) cream of tartar
150g (5oz) butter
150g (5oz) granulated sugar
175g (6oz) currants or sultanas
1 size 3 egg (beaten)
120 ml (4fl oz) milk

Preheat oven to 190°C / 375°F / Gas mark 5.

Grease and line a 15cm (6in.) round cake tin.

Sieve the flour, salt, bicarbonate of soda and cream of tartar together into a bowl. Rub the butter into the flour until it resembles fine breadcrumbs. Add the sugar and dried fruit and stir in the beaten egg and milk. Mix thoroughly. Put the mixture into the prepared tin and slightly hollow out the centre.

Bake for 15 minutes at 190°C. Then turn down to 170°C and bake for approximately 1 hour, when cooked an inserted skewer should come out clean.

Serves 8–10

K. Packer
Oxfordshire

Scrumpy Cider Cake

225g (8oz) sultanas or raisins
150ml (¼pt) dry scrumpy cider
100g (4oz) butter or margarine
100g (4oz) light or dark brown sugar
2 size 3 eggs (beaten)
225g (8oz) self-raising flour (sifted)

Preheat oven to 180°C / 350°F / Gas mark 4.

Grease tin: 18cm (7in.) square, 20.5cm (8in.) round or 1kg (2lb) loaf. Soak fruit in cider overnight.

Cream butter and sugar. Add eggs slowly, beating well.

Stir in half the flour, then the soaked fruit. Fold in remaining flour.

Bake in oven for 1 hour. Cool in tin for ½ hour and then turn out.

Serves 10   *Dorset*

Harvest Cake

450g (1lb) self-raising flour
2½ml (½ tsp) salt
225g (8oz) soft margarine
225g (8oz) caster sugar
2½ml (½ tsp) mixed spice
225g (8oz) mixed fruit
1 size 2 egg
milk to mix
sprinkling of granulated sugar

Preheat oven to 180°C / 350°F / Gas mark 4.

Grease and line a baking tin approximately 23cm × 33cm (9in. × 13in.).

In a bowl rub margarine into flour until it resembles fine breadcrumbs, add in the remaining dry ingredients. Beat the egg with the milk and add to the dry mixture. Add more milk if necessary to make a soft dropping consistency.

Pour into the prepared tin and bake for 45 minutes. The cake is ready when it springs back after being lightly touched with a finger. Cool slightly before removing from tin. Sprinkle lightly with granulated sugar and cut into squares.

Makes 16–20 squares   *E. Rowe*
                                            Essex

Orange, Chocolate and Walnut Cake

100g (4oz) soft margarine
100g (4oz) caster sugar
2 size 3 eggs
150g (5oz) self-raising flour
50g (2oz) chopped walnuts
75g (3oz) chocolate chips
grated rind and juice of a small orange
50g (2oz) soft margarine
100g (4oz) icing sugar
orange juice to taste

Preheat oven to 160°C / 325°F / Gas mark 3.

Grease and line two 18cm (7in.) sandwich tins.

Cream 100g (4oz) margarine and caster sugar together until light and fluffy. Slowly beat in the eggs, one at a time. Fold in the flour, then the nuts, chocolate and the orange rind and juice. Turn into the prepared tins and bake for 20–25 minutes.

Cream 50g (2oz) margarine with icing sugar and beat in orange juice to taste. Then sandwich the two layers together with the icing when cool.

Serves 8

R. Faunce-Brown
Somerset

Easy to Bake Chocolate Cake

CAKE
150g (5oz) self-raising flour
175g (6oz) caster sugar
175g (6oz) soft margarine
75g (3oz) drinking chocolate
3 size 3 eggs
45ml (3 tbsp) boiling water

ICING
50g (2oz) soft margarine
100g (4oz) icing sugar
25g (1 oz) drinking chocolate

Preheat oven to 180°C / 350°F / Gas mark 4.

Grease and line two 18cm (7in.) sandwich tins.

Place all the cake ingredients into a bowl and beat thoroughly for about 2 minutes. Put into the prepared tins and bake for 30–35 minutes.

Cream icing ingredients and use as filling to sandwich cakes together.

Serves 8

K. Packer
Oxfordshire

Ginger Cake

250g (8oz) plain flour
175g (6oz) brown sugar
10ml (2 tsp) ground ginger
5ml (1 tsp) baking powder
5ml (1 tsp) bicarbonate of soda
75g (3oz) margarine
30ml (2 tbsp) golden syrup
1 cup milk

Preheat oven to 180°C / 350°F / Gas mark 4.

Grease and line an 18cm × 28cm (7in. × 11in.) tin.

Mix all dry ingredients together. Rub in margarine, then add golden syrup and milk.

Mix to a doughy consistency and place in prepared tin. Bake for approximately 45 minutes.

Serves 12–16

Cambridgeshire

Sponge Parkin

250g (8oz) plain flour
5ml (1 tsp) salt
5–10ml (1–2 tsp) ground ginger according to
　　taste
75g (3oz) butter
175g (6oz) caster sugar
30ml (2 tbsp) golden syrup
1 size 3 egg
175ml (6fl oz) milk
5ml (1 tsp) bicarbonate of soda

Preheat oven to 160°C / 325°F / Gas mark 3.

Grease and line an 18cm (7in.) cake tin with baking parchment.

Sieve flour, salt and ginger together and rub fat into flour until mixture resembles breadcrumbs. Stir in sugar.

Warm syrup very slightly. Beat egg with milk and add, along with syrup, to flour and fat mixture. Mix well.

Dissolve bicarbonate of soda in a small amount of very hot water, and beat well into the mixture.

Pour into tin, and cook for 45 minutes, or until cake springs back when lightly touched with a finger.

Ideally this cake should be kept to mature 2–3 days in an airtight container.

Serves 9–12

Cambridgeshire

Yogurt Cake

Empty the contents of a 150g carton of
   plain yogurt into a bowl.
Use the carton as a measure
1 carton cooking oil
2 cartons caster sugar
3 cartons self-raising flour
5ml (1 tsp) vanilla essence
3 size 3 eggs (beaten)

Preheat oven to 160°C / 325°F / Gas mark 3.

Grease and line a 20.5cm (8in.) cake tin.

Mix all ingredients together until creamy and pour into tin.

Bake on middle shelf until firm to touch or skewer comes out clean.

This is good using a flavoured yogurt such as mandarin orange or
hazelnut, leaving out the vanilla essence.

Serves 10

S. Liddle
North Yorkshire

Cherry Madeira Cake

175g (6oz) butter
175g (6oz) caster sugar
3 size 3 eggs (beaten)
grated rind and juice of 1 orange
150g (5oz) self-raising flour
100g (4oz) plain flour
175g (6oz) glacé cherries (rinsed, dried and
   quartered)

Preheat oven to 160°C / 325°F / Gas mark 3.

Grease and line an 18cm (7in.) cake tin.

Cream butter and sugar until light and fluffy. Add eggs, beating well to incorporate.

Fold in flours alternately with orange juice, to give a stiff consistency. Then fold in cherries and orange rind.

Spoon into tin, hollowing out centre.

Bake for 1½ hours (when it is cooked an inserted skewer should come out clean).

Serves 12

I. Bryan
Warwickshire

Honey Pear Cake

225g (8oz) butter
325g (12oz) self-raising flour
100g (4oz) sultanas
225g (8oz) pears (finely chopped)
3 size 3 eggs
100g (4oz) runny honey
5ml (1 tsp) almond essence
caster sugar and juice of small lemon to
    decorate

Preheat oven to 180°C / 350°F / Gas mark 4.

Grease and line a 20.5cm (8in.) cake tin.

Rub butter into flour, and add sultanas and pears. Gently beat eggs, almond essence and honey together and stir into the first mixture until well blended.

Put into prepared cake tin and bake for 1¼ hours. An inserted skewer should pull out clean when cake is baked. While cake is still hot, sprinkle with caster sugar and then drizzle over the juice of a small lemon.

Serves 12

H. & N. Major
Gloucestershire

Snow on the Mountain Cake

50g (2oz) soft margarine
50g (2oz) caster sugar
2 size 3 egg yolks
100g (4oz) self-raising flour
pinch salt
75ml (5 tbsp) milk
5ml (1 tsp) vanilla essence

TOPPING
2 size 3 egg whites
50g (2oz) caster sugar
50g (2oz) desiccated coconut
vanilla essence

Preheat oven to 180°C / 350°F / Gas mark 4.

Grease and line an 18cm (7in.) cake tin with baking parchment.

Put all ingredients into a large bowl and beat well for 2 minutes until well mixed.

Place mixture into cake tin and smooth the top.

Prepare the topping by whisking the egg whites until stiff. Fold sugar and coconut together with a few drops of vanilla essence. Spread onto top of cake mixture.

Bake for 45–60 minutes until golden brown.

Serves 8

E. Guppy
E. Sussex

Carrot Cake

225g (8oz) sultanas
2 oranges (grated rind and juice)
225g (8oz) butter
225g (8oz) soft brown sugar
pinch of salt
4 size 3 eggs
350g (12oz) wholemeal flour
20ml (4 tsp) baking powder
10ml (2 tsp) cinnamon
225g (8oz) grated carrot

Preheat oven to 170°C / 325°F / Gas mark 3.

Grease and line a 20.5cm (8in.) cake tin.

Soak the sultanas in the orange juice for a few hours. Cream together
the butter and sugar and add the orange rind. Beat in the eggs slowly
and then fold in the flour, salt, cinnamon and baking powder. Stir in
the carrot and sultanas. Turn into the prepared tin.

Bake in the centre of the oven for approximately 2 hours until a
skewer comes out clean. Turn out and cool.

Serves 8–10

F. Sullivan
Dyfed

Rhubarb Cake

225g (8oz) self-raising flour
pinch of salt
150g (5oz) margarine
150g (5oz) caster sugar
250g (9oz) fresh rhubarb (finely diced)
75g (3oz) sultanas
2 size 3 eggs (beaten)
15ml (1 tbsp) light brown sugar

Preheat oven to 180°C / 350°F / Gas mark 4.

Grease and line a 18cm (7in.) cake tin.

Sift flour and salt into a bowl and rub in soft margarine until it resembles breadcrumbs. Stir in caster sugar, rhubarb and sultanas. Then beat in eggs.

Turn into the prepared tin, sprinkle with brown sugar and bake for 1¼ –1½ hours until well risen and golden brown. (Cake is done when an inserted skewer comes out clean.)

Leave to cool in tin for 10 minutes before removing to a wire rack to cool completely.

Serves 8–10

R. Ling
London

Sultana and Orange Cake

225g (8oz) soft margarine
225g (8oz) caster sugar
275g (10oz) self-raising flour
10ml (2 tsp) baking powder
4 size 3 eggs
30ml (2 tbsp) milk
225g (8oz) sultanas
30ml (2 tbsp) orange juice
demerara sugar

Preheat oven to 180°C / 350°F / Gas mark 4.

Grease and line an 18cm × 28cm (7in. × 11in.) tin.

Put all the ingredients except the demerara sugar in a bowl and beat until well blended. Turn into the prepared tin and smooth the top.

Bake for 25 minutes and then sprinkle the top of the cake with the demerara sugar.

Continue cooking for 25 minutes or until the cake has shrunk slightly from the sides of the tin and the top springs back when lightly touched.

Leave to cool in the tin.

Makes 18 slices

Kent

Date and Walnut Cake

1 cup dried dates (chopped)
1 cup walnuts (chopped)
1 cup brown sugar
1 cup boiling water
5ml (1 tsp) bicarbonate of soda
pinch of salt
15ml (1 tbsp) butter
2 cups plain flour
5ml (1 tsp) vanilla essence
1 size 3 egg (beaten)

Preheat oven to 180°C / 350°F / Gas mark 4.

Grease and line a 1kg (2lb) loaf tin.

In a large bowl pour the boiling water over the dates, walnuts, sugar, bicarbonate of soda and salt. Leave to stand for 15 minutes.

Rub butter into the flour. Then add to the first mixture with the beaten egg and vanilla essence and mix well.

Pour into the prepared tin and bake for 50 minutes. Leave until quite cold before eating.

Makes 12 slices

F. Cook
Merseyside

Lemon Curd Cake

175g (6oz) soft margarine
100g (4oz) caster sugar
75ml (5 tbsp) lemon curd
3 size 3 eggs
225g (8oz) self-raising flour
pinch of salt

Preheat oven to 170°C / 335°F / Gas mark 3–4.

Grease and line a 20.5cm (8in.) cake tin.

Cream the sugar and margarine together until light and fluffy. Then add the lemon curd and mix well.

Beat in the eggs one at a time with a little flour and salt. Then fold in the remaining flour. Place the mixture in the prepared tin and bake for 1 hour until lightly browned. Leave to cool in the tin.

Serves 10

S. Harmson
North Yorkshire

Coconut Cake

100g (4oz) caster sugar
100g (4oz) butter or soft margarine
100g (4oz) desiccated coconut
75g (3oz) self-raising flour
2 size 3 eggs

Preheat oven to 180°C / 350°F / Gas mark 4.

Grease and line an 18cm (7in.) cake tin.

Cream together the butter and sugar and add the eggs one at a time, beating well. Fold in the flour and coconut.

Turn into the prepared tin and bake for 30 minutes.

Serve either plain or with a glacé icing and a sprinkling of coconut on top.

Serves 8

E. Nicholson
Lancashire

Lazy Glazy Cake

100g (4oz) butter
175g (6oz) caster sugar
grated rind of 1 lemon
2 size 3 eggs (beaten)
175g (6oz) self-raising flour
pinch of salt
60ml (4 tbsp) milk
juice of 1 lemon
10ml (2 tsp) icing sugar
25g (1oz) butter
30ml (2 tbsp) brown sugar
lemon slices

Preheat oven to 180°C / 350°F / Gas mark 4.

Grease and line with baking parchment a loose-bottomed 18cm (7in.) cake tin.

Cream butter and sugar together until light and fluffy, then beat in lemon rind. Gradually beat in eggs, adding a little flour with each. Stir in milk. Fold in salt and remaining flour with metal spoon.

Pour mixture into prepared tin and bake for 1¼ hours or until golden brown and well risen (when tested, skewer should come out clean).

Mix together lemon juice and icing sugar, and pour over cake whilst still warm. Remove from tin and leave to cool on rack.

Meanwhile melt butter and sugar together over a low heat, stirring all the while. Add lemon slices and cook for about 1 minute on each side to caramelise. Remove slices and refrigerate before placing on cake.

Serves 8

I. Bryan
Warwickshire

Dream Cake Slice

PASTRY
> 175g (6oz) plain flour
> 40g (1½oz) butter or margarine
> 40g (1½oz) lard
> pinch of salt
> cold water to mix, approximately 30ml (2 tbsp)

FILLING
> 100g (4oz) butter or soft margarine
> 100g (4oz) caster sugar
> 1 size 3 egg (beaten)
> 50g (2oz) ground almonds
> 50g (2oz) ground rice
> 50g (2oz) currants
> 50g (2oz) glacé cherries (quartered)
> 30ml (2 tbsp) raspberry jam
> flaked almonds

Preheat oven to 150°C / 300°F / Gas mark 3.

Mix flour and salt in a large bowl. Rub in fats until mixture resembles breadcrumbs. Using a round-bladed knife to mix, add the cold water to form a stiff dough. Turn dough onto floured surface, knead lightly.

Roll out and line an 18cm × 28cm (7in. × 11in.) tin.

Cream butter and sugar and beat in egg. Fold in ground almonds and ground rice, then cherries and currants.

Spread base of pastry with raspberry jam and then pour filling evenly over it. Sprinkle with a few flaked almonds.

Bake for 1¼ hours until golden brown. Cool.

Makes 16 slices   *Northumberland*

Citrus Slices

175g (6oz) butter
175g (6oz) caster sugar
3 size 3 eggs (beaten)
175 (6oz) self-raising flour
grated rind of 1 lemon
grated rind of 1 orange
3 drops vanilla essence
juice of ½ lemon
juice of ½ orange
icing sugar

Preheat oven to 190°C / 375°F / Gas mark 5.

Grease and line an 18cm × 28cm (7in. × 11in.) tin.

Cream together the butter and sugar. Gradually work in the beaten eggs, alternately with the flour. Add the grated fruit rind and the essence, and pour into the prepared tin. Bake for 30 minutes.

Cool and ice with the fruit juices mixed with enough icing sugar to make a spreading consistency.

Cut into slices and serve.

Makes 16 slices   *Pembrokeshire*

Matilda Slice

175g (6oz) butter
20ml (2 dsp) golden syrup
100g (4oz) granulated sugar
100g (4oz) sultanas
50g (2oz) glacé cherries (chopped)
350g (12oz) self-raising flour

Preheat oven to 180°C / 350°F / Gas mark 4.

Grease and line an 18cm × 28cm (7in. × 11in.) tin.

Melt together the butter and syrup in a saucepan. Add the sugar, sultanas and cherries. Sieve flour into a bowl, add the melted mixture and beat well.

Turn into the prepared tin and bake for 20 minutes. Leave to cool in the tin.

Makes 16 slices

K. Packer
Oxfordshire

Banana Coconut Fingers

CAKE
100g (4oz) soft margarine
225g (8oz) self-raising flour
100g (4oz) caster sugar
2 size 3 eggs (beaten)
60ml (4 tbsp) milk
2 ripe medium bananas (mashed)

ICING
100g (4oz) soft margarine
175g (6oz) icing sugar (sieved)
25g (1oz) desiccated coconut

Preheat oven to 180°C / 350°F / Gas mark 4.

Grease and line an 18cm × 28cm (7in. × 11in.) tin.

In a large bowl mix together all the cake ingredients and beat well for at least 2 minutes. Spoon into the prepared tin and bake for 35 minutes.

To make icing, cream the margarine and icing sugar until light and fluffy and then add the coconut. Ice cake when it is cool.

Makes 16 slices

D. Whitehall
Buckinghamshire

Fruit Slices

PASTRY

225g (8oz) plain flour
50g (2oz) butter or hard margarine
50g (2oz) lard
pinch of salt
cold water to mix

FILLING

75g (3oz) sultanas
75g (3oz) currants
2½ml (½ tsp) cinnamon
2½ml (½ tsp) ground cloves
30ml (2 tbsp) melted butter
50g (2oz) light soft brown sugar
caster sugar (for dredging)

Preheat oven to 200°C / 400°F / Gas mark 6.

Mix the flour and salt in a large bowl, and rub in the fats until mixture resembles fine breadcrumbs. Using a round-bladed knife to mix, add the cold water to form a stiff dough. Turn dough out onto a floured surface, knead lightly and divide pastry into two equal quantities. Roll each piece into an 18cm (7in.) square that is 5mm (¼in.) thick and place onto a baking sheet.

Mix the fruit, spices and melted butter with the sugar and spread over one square to within 1cm (½in.) of the edge. Brush the edges with cold water and lay the other pastry square on top. Press down lightly and brush the top with milk. Bake for approximately 20 minutes until golden brown. Remove from oven and dredge with caster sugar.

Makes 16 portions

Warwickshire

Teabreads and Biscuits

Malt Loaf

450g (1lb) self-raising flour
25g (1oz) butter
15ml (1 tbsp) black treacle
15ml (1 tbsp) malt extract
175g (6oz) light brown soft sugar
100g (4oz) raisins
350ml (12fl oz) warmed milk

Preheat oven to 150°C / 300°F / Gas mark 2.

Grease and line the base of 2 ×1lb loaf tins.

Place all the ingredients into a mixing bowl and mix together well.
Put in the prepared tins and bake for 1 hour.

Serves 12

E. Nicholson
Lancashire

All-Bran Fruit Loaf

1 heaped teacup of All-Bran
1 heaped teacup of dried mixed fruit
1 teacup of demerara sugar
1 teacup of milk
1 heaped teacup of self-raising flour

Preheat oven to 180°C / 350°F / Gas mark 4.

Grease and line the base of a 1kg (2lb) loaf tin.

Place All-Bran, sugar and fruit into a bowl and mix well. Stir in the milk and leave to stand for an hour. Then add the sieved self-raising flour, mixing well.

Put into the prepared tin and bake for 1½ hours. Leave to cool in the tin.

Serve sliced with butter.

Serves 12

F. Borey
Hampshire

Bara Brith

500g (18oz) dried mixed fruit
1 mug soft brown sugar
1 mug strong black tea
2 mugs self-raising flour
10ml (2 tsp) cinnamon
5ml (1 tsp) nutmeg
2 size 3 eggs (beaten)

Preheat oven to 150°C / 300°F / Gas mark 2.

Grease and line two ½kg (1lb) loaf tins.

Mix the fruit, sugar and tea in a large bowl and leave to soak overnight.

Sieve the flour and spices together into the soaked mixture, then add the eggs.

Divide the mixture between the two loaf tins and smooth the tops flat. Bake for 2 hours.

Turn out and cool. Serve cut into slices and buttered.

Makes 12 slices per loaf

F. Sullivan
Dyfed

Harvest Fruit Loaf

100g (4oz) sugar
100g (4oz) margarine
2 size 3 eggs
150ml (¼pt) milk
350g (12oz) mixed fruit
225g (8oz) self-raising flour
5ml (1 tsp) mixed spice
walnuts and cherries to decorate

Preheat oven to 160°C / 325°F / Gas mark 3.

Place all ingredients except walnuts and cherries in a bowl and mix well. Put into a ½kg (1lb) loaf tin, decorate with walnuts and cherries and then bake for 1½ to 1¾ hours.

Leave in tin for 10 minutes after taking out of oven. Then turn out onto wire rack to cool.

Makes 12 slices

J. Todd
E. Yorkshire

Banana Loaf

100g (4oz) soft margarine
100g (4oz) soft brown sugar
225g (8oz) self-raising flour
2½ml (½ tsp) bicarbonate of soda
pinch of salt
2 size 3 eggs (beaten)
4 ripe bananas
15ml (1 tbsp) lemon juice
50g (2oz) dates (chopped)
50g (2oz) walnuts (chopped)

Preheat oven to 160°C / 325°F / Gas mark 3.

Grease and line a 1kg (2lb) loaf tin.

Cream together the margarine and the sugar, then beat in the flour, salt and bicarbonate of soda with the eggs. Mash the bananas with the lemon juice and add to the mixture with the dates and walnuts.

Spoon into the prepared loaf tin and bake for 1¼ hours.

Cool in the tin.

Serve sliced with butter if desired.

Makes 12 slices

H. & N. Major
Gloucestershire

Welsh Cakes

225g (8oz) self-raising flour
pinch of salt
75g (3oz) margarine
75g (3oz) caster sugar
50g (2oz) currants
1 size 3 beaten egg
little cold water to mix

Sift flour and salt into a basin, and rub in the margarine until mixture is like breadcrumbs. Add the sugar, currants and the beaten egg and mix to the consistency of a moist pastry dough with a little water. Roll out on a floured board to approximately 0.5cm (¼in.) thick, and cut out rounds with a 6.5cm (2½in.) cutter.

Heat a griddle or heavy frying pan and grease it lightly with oil. Test the heat of the griddle by cooking a single cake; if it's too hot, the cakes burn before the inside is cooked. Cook cakes on both sides until just golden. Grease pan between batches.

Put onto a wire rack to cool. Store in an airtight container.

Serve buttered. Lovely warmed.

Makes 20 cakes

Dyfed

Spiced Harvest Cakes

15ml (1 tbsp) dried yeast
75g (3oz) butter
225ml (7½ fl oz) milk
75g (3oz) brown sugar
grated rind of ½ lemon
550g (1¼ lb) plain flour
pinch of salt
5ml (1 tsp) mixed spice
2 size 3 eggs (beaten)
175g (6oz) mixed dried fruit

Preheat oven to 220°C / 425°F / Gas mark 7.

Make up yeast according to the instructions on the packet. Adjust liquid accordingly.

Melt the butter gently with the milk, sugar and lemon rind in a small saucepan. Do not boil!

Sieve the flour with the salt and mixed spice into a large bowl. Make a well in the centre and stir in the milk and sugar mixture. Then add the yeast and beaten eggs and stir well. Cover the bowl with cling film and leave to rise in a warm place for 1 hour.

When the mixture is well risen, turn out onto a floured board and add the mixed fruit to the dough, kneading for 5 minutes.

Divide the dough into 16 pieces and place on a lightly buttered tin. Leave to rise for another hour, brush with beaten egg or milk, and bake for 10–15 minutes.

Makes 16 cakes

M. Jackson
Norfolk

Sidney Specials

175g (6oz) margarine
100g (4oz) soft brown sugar
25g (1oz) Weetabix, crumbled
150g (5oz) self-raising flour
50g (2oz) desiccated coconut
15ml (1 tbsp) cocoa powder
100g (4oz) cooking chocolate for the topping

Preheat oven to 160°C / 325°F / Gas mark 3.

Grease and line an 18cm × 28cm (7in. × 11in.) tin.

Melt the margarine in a medium saucepan. Add the sugar and dissolve; do not boil. Allow to cool slightly and mix in all the other ingredients except chocolate.

Spread into prepared tin and bake for 30 minutes. Remove from oven. Melt the chocolate and spread over the top.

Cut into fingers whilst still warm, leaving in the tin to cool.

Makes 16 squares

M. Hardy
Hertfordshire

Oatmeal Scones

175g (6oz) self-raising flour
100g (4oz) wholemeal flour
100g (4oz) butter
150g (5oz) medium oatmeal
150ml (5fl oz) milk
1 size 3 egg (beaten)
pinch of salt

Preheat oven to 200°C / 400°F / Gas mark 6.

In a bowl, rub butter into the flours and oatmeal until mixture looks like breadcrumbs. Mix in the egg and milk with a wooden spoon – this makes a sticky dough.

Shape the whole of the dough in a round approximately 20.5cm (8in.) in diameter on a lightly greased baking tray. Mark the top into eight wedges.

Bake for 30 minutes.

Makes 8 wedges

F. Cook
Merseyside

Chocolate Crunch

100g (4oz) margarine
10ml (1 dsp) caster sugar
15ml (1 tbsp) golden syrup
22½ml (1½ tbsp) drinking chocolate
50g (2oz) sultanas
225g (8oz) broken digestive biscuits
100g (4oz) cooking chocolate

Grease and line the base of an 18cm (7in.) square cake tin.

Melt the margarine in a pan. Stir in the syrup, sugar, drinking chocolate and sultanas, and then add the broken biscuits.

Place the biscuit mixture in the tin and press down firmly. Spread with the melted chocolate and leave to set.

Then cut into squares.

Makes 16 squares

S. Harmson
North Yorkshire

ℛaspberry Buns

100g (4oz) soft margarine or butter
225 (8oz) self-raising flour
100g (4oz) caster sugar
1 size 3 egg
vanilla essence
raspberry jam

Preheat oven to 180°C / 350°F / Gas mark 4.

Rub margarine into flour, stir in sugar, mix in egg and vanilla essence. The mixture will resemble a soft pastry.

Divide mixture into small balls. Flatten slightly and make a small well in the centre. Place a little jam in the well and then close partially.

Cook on a lightly greased baking tray for approximately 15 minutes.

Makes 16 buns

I. Thomas
Glamorgan

Kensall Shortbread

100g (4oz) caster sugar
225g (8oz) butter
225g (8oz) plain flour
100g (4oz) cornflour or ground rice

Preheat oven to 180°C / 350°F / Gas mark 4.

On a board or work surface, rub the sugar into the butter until very soft. Work in the flour and the cornflour or ground rice.

Roll out to 0.5cm (¼in.) thick and put into circles using a 6.5cm (2½in.) cutter. Prick with a fork. Place on a greased baking sheet and bake for 20 minutes until a pale golden colour.

Leave to cool on baking sheet.

Makes 24 biscuits

Ruth Mills
Somerset

Ginger Snaps (Wartime Recipe)

75g (3oz) margarine
250g (9oz) plain flour
100g (4oz) caster sugar
30ml (2 tbsp) golden syrup
7½ml (¾ tsp) bicarbonate of soda

Preheat oven to 180°C / 350°F / Gas mark 4.

Rub the margarine into the flour, then add the sugar and the bicarbonate of soda and mix well. Stir in the golden syrup and mix to form a stiff paste.

Roll out thinly and cut into rounds using a 5cm (2in.) biscuit cutter.
Place on a greased baking sheet and cook for 15–20 minutes.

Makes about 50 biscuits

M. Phethean
Suffolk

Walnut Biscuits

100g (4oz) butter
75g (3oz) caster sugar
100g (4oz) plain flour
25g (1oz) semolina
25g (1oz) walnuts (chopped)

Preheat oven to 190°C / 375°F / Gas mark 5.

Cream the butter and sugar together until light and fluffy. Add the
flour, semolina and nuts, and stir into a stiff dough.

Form into a roll 18–20.5cm (7in.–8in.) long. Cut into approximately 16
rounds. Place on a greased baking sheet and cook for 15–20 minutes.
Leave to cool on the baking sheet.

Makes 16

E. Southam
Cambridgeshire

Priory Biscuits

175g (6oz) rolled oats
175g (6oz) self-raising flour
225g (8oz) caster sugar
100g (4oz) butter (melted)
15ml (1 tbsp) treacle
5ml (1 tsp) bicarbonate of soda
45ml (3 tbsp) hot water

Preheat oven to 190°C / 375°F / Gas mark 5.

Mix oats, flour, sugar and bicarbonate of soda in a large bowl. Stir in the melted butter and treacle, then add the hot water. Drop teaspoons of the mixture onto a greased baking tray.

Bake for 10–15 minutes.

Makes about 16 biscuits

South Yorkshire

Devonshire Biscuits

350g (12oz) self-raising flour
150g (5oz) butter
150g (5oz) caster sugar
1 size 3 egg

Preheat oven to 180°C / 350°F / Gas mark 4.

Rub the fat into the flour until it resembles breadcrumbs. Stir in the sugar, add the egg and mix to a stiff dough.

Roll out thinly on a floured board, and cut out biscuits using a 6.5cm (2½in.) cutter.

Place on a baking sheet and bake for 10–15 minutes.

Makes about 40 biscuits

Devon

Flapjacks

50g (2oz) rolled oats
50g (2oz) self-raising flour
75g (3oz) crushed cornflakes
100g (4oz) caster sugar
100g (4oz) butter or hard margarine
15ml (1 tbsp) golden syrup
grated rind of ½ lemon

Preheat oven to 180°C / 350°F / Gas mark 4.

Grease and line an 18cm × 28cm (7in. × 11in.) tin.

Put all the dry ingredients into a bowl and mix well. Melt the butter and syrup over a low heat, pour over the dry ingredients and stir thoroughly.

Turn into the prepared tin and flatten down evenly.

Bake for approximately 15–20 minutes, taking care not to overcook.

Makes 12 squares

M. Phethean
Suffolk

Golden Shorties

175g (6oz) soft margarine
50g (2oz) icing sugar
175g (6oz) plain flour
almond or vanilla essence

Preheat oven 180°C / 350°F / Gas mark 4.

Mix all the ingredients together. Place the mixture into a piping bag
and pipe onto a lightly greased baking tray. Cook for 15 minutes until
golden brown. Allow to cool, and store in an airtight tin.

Makes 15–20 biscuits

Clwyd

Melting Moments

100g (4oz) lard
50g (2oz) caster sugar
175g (6oz) self-raising flour
1 size 3 egg
5ml (1 tsp) vanilla essence
rolled oats

Preheat oven to 180°C / 350°F / Gas mark 4.

Cream lard and sugar together. Add the flour and then the egg and
vanilla essence to make a sticky mixture.

Take teaspoonfuls of the mixture and make into balls, then roll each
ball in the rolled oats.

The mixture is less likely to stick to the hands if hands are rinsed in
cold water before rolling.

Place balls on a lightly greased baking tray and bake for 10–15 minutes. Leave to cool before removing from tray.

Makes 30 biscuits

N. J. Hickin
Hampshire

Oat Cookies

100g (4oz) margarine
75g (3oz) caster sugar
1 size 3 egg
100g (4oz) self-raising flour
50g (2oz) medium oatmeal
rolled oats
glacé cherries to decorate

Preheat oven to 160°C / 325°F / Gas mark 3.

Cream fat and sugar together, and beat in the egg.

Stir in the flour and the oatmeal. Take teaspoonfuls of the mixture and form into balls and roll them in the rolled oats. Place on a greased baking sheet.

Place a cherry quarter in the centre of each ball.

Bake for 20 minutes or until a pale golden colour.

Leave to cool on baking sheet.

Makes 18 biscuits

V. Story
Nottinghamshire

Orange Biscuits

225g (8oz) self-raising flour
150g (5oz) margarine or butter
150g (5oz) caster sugar
grated rind of 2 oranges
1 size 3 egg (separated)
15ml (1 tbsp) milk
sprinkling of caster sugar for topping

Preheat oven to 180°C / 350°F / Gas mark 4.

Lightly grease 2 baking sheets.

Place flour in a bowl and rub in margarine or butter until mixture resembles breadcrumbs. Mix in caster sugar and orange rind.

Beat egg yolk and milk together, and add to the dry ingredients in the bowl. Mix to form a stiff dough.

Turn out onto a floured board and knead until smooth. Roll out to 0.25cm (⅛in.) thickness.

In a small bowl beat the egg white lightly with a fork. Brush over the rolled out dough and sprinkle lightly with caster sugar.

Cut into rounds with a 5cm (2in.) fluted cutter. Knead together trimmings and roll out again. Repeat brushing with egg white and sprinkling with sugar.

Place biscuits on baking sheets, allowing room for spreading.

Bake for 10–15 minutes until a pale golden-brown colour. Cool on a wire rack.

Makes 40 biscuits

P Sorensen
Devon

Preserves
and
Chutneys

Rhubarb and Ginger Jam

1.35kg (3lb) rhubarb
1.35kg (3lb) granulated sugar
grated rind and juice of 3 lemons
25g (1oz) bruised dried root ginger or 5cm
    (2in.) fresh ginger
100g (4oz) finely chopped crystallised ginger or
    ginger in syrup

Wipe the rhubarb and cut it into small chunks. Place in a bowl and sprinkle the sugar on in layers. Add the lemon juice and leave to stand overnight.

The next day put into a preserving pan. Add the lemon rind, tie the root ginger in a muslin bag and add to the pan. Heat gently and ensure all sugar is dissolved. Then turn up the heat and bring to the boil, boiling rapidly until setting point is reached.

To test whether jam has reached setting point after about 10 minutes of boiling, place a teaspoonful of jam on a cold saucer, and if it wrinkles when pushed with a finger it is ready. If not, return to the boil and test again after about 5–10 minutes. When setting point is reached, stir in the finely chopped ginger.

Pot into hot sterilised jars and seal.

Makes 6 x 450g (1lb) jars

H. & N. Major
Gloucestershire

Crab Apple Jelly

crab apples
water, 450ml (¾pt) to 450g (1lb) apples
sugar, 450g (1lb) to 600ml (1pt) juice

Wash and cut up crab apples. Put in water and bring to the boil. Simmer until fruit is soft. Strain through a jelly bag or muslin-lined sieve.

Measure the juice into a heavy saucepan and add sugar. Bring to the boil and boil rapidly for 5 minutes. Then test for setting in the usual way (if a drop of jelly on a cooled saucer crinkles when gently pushed, it has reached setting point). If it has not reached setting point, boil for another 5 minutes and then test again.

Put in small, hot sterilised jars. Seal.

Lovely with cold pork.

D. Edwards
Hereford & Worcester

Apple Ginger Jam

1.35kg (3lb) apples
600ml (1pt) water
5ml (1 tsp) ground ginger
grated rind and juice of 2 lemons
1.35kg (3lb) granulated sugar
100g (4oz) crystallised ginger

Peel, core and chop the apples. Put the peel and cores in a muslin bag.

Lightly butter a preserving pan and place in it the chopped apples and water, together with the lemon juice, grated rind and ground ginger. Hang the muslin bag in the pan. Cook the mixture until tender.

Remove the bag of peel and squeeze out all the juices. Add the sugar and chopped ginger, and over low heat slowly dissolve the sugar. When sugar has dissolved, boil rapidly until setting point is reached.

Check after 10–15 minutes for a set. Place a teaspoonful of jam on a cold saucer; if it wrinkles when pushed with a finger, it is ready. If it isn't, boil for a few minutes more, then test again.

Pot into warmed sterilised jars and seal.

Makes 6 × 450g (1lb) jars

Pumpkin and Grapefruit Marmalade

1.12kg (2½lb) pumpkin flesh
2½ grapefruit
900g (2lb) sugar
juice of 1 lemon
knob of butter

Remove the rind of the grapefruit with a zester. Mince or finely chop the grapefruit flesh and the pumpkin. Place in a bowl and sprinkle with a little of the sugar. Leave overnight to draw out the juices.

Lightly butter a preserving pan. Place the grapefruit and pumpkin mixture in the pan with the peel, and cook until soft. Stir in the sugar over a low heat until dissolved. Bring to the boil, and after 10 minutes test for setting point by placing a drop of the marmalade on a cool saucer. If it crinkles when gently pushed, it has reached setting point. If setting point is not reached, boil again and test after 10 minutes. Put in hot sterilised jars.

Makes 3 x 350g (12oz) jars

D. Quaife
East Sussex

Chunky Cottage Marmalade

900g (2lb) Seville oranges
1 sweet orange
1 lemon
1.57kg (3½lb) granulated sugar

Wash fruits and place whole in a large saucepan. Cover with water, bring to the boil and simmer gently with a lid on for 2 hours or until fruit is tender.

When ready, remove fruit from liquid with a knife and fork, and cut each orange in half and scrape out pulp, discarding any pips.

Cut rinds into chunks and place with the pulp into a lightly buttered preserving pan (reduces scum). Add 300ml (½ pt) of strained liquid together with the sugar and heat gently until sugar has dissolved.

Boil briskly for 12 minutes and then start testing for a set. This is done by removing pan from heat and placing a spoonful of marmalade on a cooled saucer. The surface will crinkle when pushed with a finger. If it does not, return to the heat, gently boil for a further 5 minutes and test again.

Pot up in warmed jars, cover and label.

Makes 6 x 450g (1lb) jars

West Midlands

Dried Apricot Jam

675g (1½lb) dried apricots (chopped)
1.8 litres (3pt) water
1.35kg (3lb) granulated sugar
100g (4oz) almond flakes

Soak chopped apricots in the water for 48 hours.

Lightly butter a preserving pan (this helps to reduce scum), bring the apricots and water to the boil, reduce heat and simmer gently until tender. Add granulated sugar and almonds over low heat. When sugar has dissolved, boil gently for 20 minutes.

To test for a set, place a small spoonful of jam on a saucer. It should crinkle when pushed gently with a finger. If set is not reached, boil for another 5–10 minutes and test again.

Pot up into clean, warmed jars and label.

Makes 5 x 450g (1lb) jars

Somerset

Orange Curd

rind of 3 oranges
juice of 2 oranges
225g (8oz) granulated sugar
100g (4oz) butter
2 size 3 eggs (beaten)

Put rind, juice, sugar and butter into a bowl over a pan of hot (not boiling) water. Cook until butter has melted. Add eggs and stir in well.

Continue cooking, stirring occasionally, until mixture thickens enough to coat the back of a wooden spoon.

Pour into hot jars. Seal.

Makes 2 x 350g (12oz) jars

M. Prothero
Hereford & Worcester

Pickled Walnuts

Use green walnuts, which have to be picked before 15 July before the hard shell has formed inside.

Make a brine by boiling 100g (4oz) salt in 1.2 litres (2pt) water and allow to cool. Make as much as is needed to cover the walnuts.

Place walnuts in brine for 6 days, stirring daily. After 6 days drain and cover with fresh brine for another 3 days. Drain. Spread the nuts out on trays and place in the sun to turn black. When they have turned black, place walnuts in clean jars and cover with pickling vinegar.

Seal jars. Store in a cool place and they will keep for several months.

E. Downes
Norfolk

Rhubarb Chutney

900g (2lb) rhubarb (chopped)
450g (1lb) sultanas
2 large apples (peeled, cored and finely
    chopped)
1 large onion (finely chopped)
25g (1oz) ground ginger
25g (1oz) salt
600ml (1pt) spiced vinegar
900g (2lb) light brown soft sugar

Put all ingredients into a preserving pan. Bring to the boil slowly, stirring well. Reduce heat and simmer until chutney has thickened and a wooden spoon can be drawn cleanly across the bottom of the pan. Pour into warmed jars, seal and store.

Makes 3 × 450g (1lb) jars   *Dorset*

Salad Cream

350g (12oz) granulated sugar
4 size 3 eggs
600ml (1pt) fresh milk
15ml (1 tbsp) salt
10ml (1 dsp) dry mustard
10ml (1 dsp) cornflour
300ml (½pt) wine vinegar

In a double saucepan, mix together all ingredients, adding the vinegar last (otherwise it may curdle). Cook for 1 hour, stirring occasionally.

This will keep for several months in a screw-top jar.

E. Sowden
Devon

Apple Chutney

2kg (5lb) apples (peeled, cored and finely
    chopped)
900g (2lb) onions (finely chopped)
900g (2lb) soft light brown sugar
5ml (1 tsp) mustard powder
5ml (1 tsp) ground ginger
450g (1lb) sultanas
600ml (1pt) malt vinegar

Place all ingredients into a large saucepan, bring to boil and simmer
for approximately 2 hours, stirring occasionally.

Chutney is ready when nearly all the vinegar is absorbed, and mixture
has thickened to a soft consistency.

Put in warm jars, and seal.

Makes 8 x 450g (1lb) jars

I. Thomas
Glamorgan

Plum Chutney

2.25kg (5lb) plums (stoned)
900g (2lb) onions (chopped)
1.8kg (4lb) apples (peeled and chopped)
750ml (1¼ pt) white vinegar
15ml (1 tbsp) allspice in muslin bag
15ml (1 tbsp) ground ginger
15ml (1 tbsp) ground cloves
30ml (2 tbsp) salt
900g (2lb) granulated sugar
900g (2lb) dark brown sugar

Put all ingredients except sugars in a preserving pan.

Bring to boil and simmer gently until fruits and onion are soft. Then add sugar and stir until it is dissolved.

Bring to brisk rolling boil and boil for 20 minutes, stirring all the time (it catches quickly).

Pot up and seal.

Makes about 12 x 450g (1lb) jars

M. Phethean
Suffolk

Beetroot Chutney

900g (2lb) beetroot (uncooked)
600ml (1 pt) malt vinegar
5ml (1 tsp) salt
1 ml (¼ tsp) ground ginger
450g (1 lb) apples (peeled and finely chopped)
2 large onions (finely chopped)
225g (8oz) dark brown sugar

Cook beetroot whole in saucepan of salted water. When cooked, cool, peel and then finely chop.

In a large pan place vinegar, salt and ground ginger, add apples and onions. Bring to the boil and cook for 20 minutes. Add sugar and stir until dissolved. Then add beetroot and boil for 10 minutes.

Remove from heat and bottle.

Makes about 5 x 350g (12oz) jars

G. Jones
Gwynedd

Runner Bean Chutney

900g (2lb) runner beans (trimmed and sliced)
675g (1½lb) chopped onions
900ml (1½ pt) white vinegar
900g (2lb) demerara sugar
22½ml (1½ tbsp) cornflour
22½ml (1½ tbsp) turmeric
pinch of salt
5ml (1 tsp) dry mustard

Cook sliced beans in salted water until tender (but with a slight bite).
Then drain. Boil onions in 300ml (½ pt) of the vinegar.

Mix sugar, cornflour, turmeric and mustard powder to a smooth paste
with a little vinegar.

Add rest of vinegar and the beans to the onions and boil for
15 minutes.

Add the paste to the bean and onion mixture and boil for 10 minutes.
When cool, pot and seal.

Makes 6 × 450g (1lb) jars

G. Finnemore
Cornwall

Christmas

Freezer Mincemeat

450g (1lb) cooking apples
150ml (¼pt) cider
450g (1lb) raisins
100g (4oz) cut mixed peel
175g (6oz) shredded suet
450g (1lb) sultanas
350g (12oz) currants
450g (1lb) soft brown sugar
2½ml (½ level tsp) mixed spice
rind and juice of 2 lemons
45ml (3 tbsp) rum or brandy

Peel, core and cook apples in cider until soft and pulpy. Put aside to cool.

Place raisins and mixed peel into a large bowl, together with suet and remaining dried fruit.

When apples are cool, stir in the sugar, spice, juice and rind of lemons. Add the apple mixture to the dried fruit mixture.

Add the brandy or rum and stir well.

Put into container and leave for 3 or 4 days in refrigerator to mature and allow fruit to soak up the juices.

Transfer into smaller containers (500g margarine pots are ideal) and freeze until required. Thaw at room temperature before use.

Makes 5 x 500g containers
Yield 5lb

Cornwall

Lemon Mincemeat

juice and rind of 2 large lemons
6 Bramley apples (peeled and cored)
225g (8oz) suet
450g (1lb) currants
225g (8oz) granulated sugar
75g (3oz) mixed peel
10ml (2 tsp) mixed spice

With a lemon zester, remove peel from lemons. Blanch in boiling water for 5 minutes. In a food processor chop the peeled and cored apples and the lemon peel.

Place all the remaining ingredients in a large bowl with the minced apple mixture and stir well.

Place in clean, warmed jam jars and seal.

Makes 3 x 450g (1lb) jars

B. King
Somerset

Mince Pies

100g (4oz) plain flour
25g (1oz) butter or margarine
25g (1oz) lard
pinch of salt
cold water, to mix

FILLING
12 × 5ml (1 tsp) mincemeat

TOPPING
25g (1oz) caster sugar
100g (4oz) butter
1 size 3 egg yolk
100g (4oz) self-raising flour

Preheat oven to 200°C / 400°F / Gas mark 6.

Mix flour and salt in a large bowl, then rub in fats until mixture resembles breadcrumbs.

Using a round-bladed knife to mix, add the cold water to form a stiff dough.

Turn dough onto a floured surface and knead lightly.

Roll out and line a 12 hole bun tin.

Put 5ml (1 tsp) of mincemeat in each pastry case.

Cream butter and sugar until light and fluffy, beat in egg yolk and then add the flour.

Put the mixture into a piping bag fitted with a no. 2 star tube. Pipe a ring on each tart, leaving a small hole in the centre.

Bake approximately 10–15 minutes until golden brown.

Makes 12

M. Adams
Powys

Christmas Cake

450g (1lb) dried mixed fruit
juice and rind of 1 orange
15ml (1 tbsp) brandy
100g (4oz) butter
100g (4oz) dark brown sugar
15ml (1 tbsp) black treacle
3 size 3 eggs
175g (6oz) plain flour
2½ml (½ tsp) ground cinnamon
50g (2oz) ground almonds
50g (2oz) glacé cherries

Preheat oven to 140°C / 275°F / Gas mark 1.

Grease and line an 18cm (7in.) cake tin.

To plump up the fruit, soak it with the orange juice and rind and brandy overnight in a bowl covered with a tea towel.

Cream butter and sugar together until soft. Add the treacle, then beat in the eggs. Fold in the flour, cinnamon and ground almonds, then add the cherries and soaked fruit mixture and stir well.

Put mixture into the prepared tin and cook for approximately 3 hours. When the cake is done it should shrink slightly from the sides of the tin, and an inserted skewer should pull out clean.

Cool before removing from the tin.

Serves 10–12   *Dyfed*

Christmas Cake

275g (10oz) butter
275g (10oz) soft brown sugar
grated rind of 1 lemon
22½ml (1½ tbsp) black treacle
6 size 3 eggs
300g (11oz) plain flour
7½ml (1½ tsp) mixed spice
2½ml (½ tsp) nutmeg
100g (4oz) glacé cherries
100g (4oz) cut mixed peel
450g (1lb) currants
275g (10oz) sultanas
175g (6oz) raisins
100g (4oz) chopped walnuts
45–60ml (3–4 tbsp) brandy

Preheat oven to 150°C / 300°F / Gas mark 2.

Grease and line a 25.5cm (10in.) round cake tin.

In a large bowl cream together butter and sugar until light and fluffy. Add lemon rind, then treacle, combining well. Thoroughly beat in the eggs one at a time. Fold in the flour, spices, then add all the dried fruit, cherries, mixed peel and walnuts, mixing well.

Turn into the prepared cake tin and wrap a layer of brown paper, tied with string, around the outside of the tin for extra protection.

Bake for approximately 4 hours (check after 3 hours). When cooked an inserted skewer should come out cleanly. Leave to cool in tin. Pierce cake with a skewer and pour brandy over it. When cold remove from tin.

Serves 16–20

Christmas Pudding

900g (2lb) breadcumbs
half a nutmeg, grated
450g (1lb) currants
450g (1lb) soft brown sugar
2 size 3 eggs (beaten)
450g (1lb) grated carrots
2½ml (½ tsp) mixed spice
2½ml (½ tsp) salt
150ml (¼pt) rum
grated rind of 1 lemon

This pudding may be made two weeks in advance of Christmas.

Mix all the ingredients together in a large bowl, cover with a cloth and leave overnight to enable flavours to develop.

Next day place the mixture into a lightly greased 1.2 litre (2pt) basin. Cover with a sheet of greaseproof paper and foil, tied lightly with string. For ease of removing basin from steamer make a string handle.

Place the pudding in a large steamer or saucepan, and steam over simmering water for 8 hours, making sure that the pan does not boil dry. Let the pudding get cold and then replace with fresh greaseproof paper and foil.

When pudding is reheated it must be steamed or boiled for 3 hours.

Serves 8

N. Fenning
Suffolk

Christmas Pudding

100g (4oz) shredded suet
50g (2oz) self-raising flour
100g (4oz) white breadcrumbs
10ml (2 tsp) mixed spice
7½ml (1½ tsp) grated nutmeg
2½ml (½ tsp) ground cinnamon
225g (8oz) dark brown soft sugar
100g (4oz) sultanas
100g (4oz) raisins
275g (10oz) currants
50g (2oz) mixed peel
1 large cooking apple (chopped)
zest and juice of 1 orange
zest and juice of 1 lemon
30ml (2 tbsp) rum, brandy or sherry
150ml (5fl oz) stout
2 size 3 eggs

Mix the pudding a day before steaming, so flavours can develop. May be made a month before Christmas.

Place suet, flour, breadcrumbs, spices and sugar into a large bowl. Stir well, add dried fruit, mixed peel, chopped apple and the orange and lemon zest and juice.

In a small bowl put the rum, brandy or sherry, the stout and the eggs and beat well. Add to the suet mixture and stir well – don't forget the Christmas wishes! Cover with a cloth and leave in a cool place overnight.

The next day spoon the mixture into a lightly greased 1.2 litre (2pt) basin. Cover with a sheet of greaseproof paper and a sheet of foil tied tightly with string. Use more string to make a handle.

In a large steamer or saucepan, steam the pudding over or in simmering water for 8 hours, checking water levels regularly so that the pan does not boil dry.

When done let pudding get quite cold before removing, and replacing with fresh greaseproof paper, foil and string.

On Christmas Day, steam as before for 3–4 hours.

Serves 8–10

E. Downes
Norfolk

Turkey and Tomato Casserole

1 medium onion (finely chopped)
3 rashers back bacon (chopped)
100g (4oz) mushrooms (sliced)
450g (1lb) cooked cubed turkey
2½ml (½ tsp) dried mixed herbs
150ml (¼pt) wine (red or white)
2 sticks celery (chopped)
411g (14½oz) tin tomatoes
150ml (¼pt) chicken stock
salt and pepper

Preheat oven to 180°C / 350°F / Gas mark 4.

Fry and lightly brown onion, bacon and mushrooms. Place in a medium-sized casserole with all other ingredients and season to taste. Cook for 45 minutes.

Serves 4

E. Stanton
Northamptonshire

Cold Turkey Leftovers

900g (2lb) cold cubed turkey
15–30ml (1–2 tbsp) honey
60ml (4 tbsp) mixed chutney
10–15ml (½–1 tbsp) curry powder to taste
small glass white wine or cider
300ml (½pt) double cream
300ml (½pt) mayonnaise

Mix together all ingredients except the turkey. Then incorporate bite-sized turkey cubes into the sauce.

Refrigerate at least 2 hrs before serving to mature the flavours. Serve with salad and crusty bread.

Serves 8–12

J. Smith
Berkshire

Index